GOVERNING NATURE AND THE MAKING OF WORLD ORDER

Edited by
Elana Wilson Rowe, Paul Beaumont
and Lucas de Oliveira Paes

First published in Great Britain in 2025 by

Bristol University Press
University of Bristol
1–9 Old Park Hill
Bristol
BS2 8BB
UK
t: +44 (0)117 374 6645
e: bup-info@bristol.ac.uk

Details of international sales and distribution partners are available at bristoluniversitypress.co.uk

© Editorial selection and editorial matter © 2025 Elana Wilson Rowe,
© 2025 Paul Beaumont, © 2025 Lucas de Oliveira Paes

The digital PDF and ePub versions of this title are available open access and distributed under the terms of the Creative Commons Attribution-NonCommercial-NoDerivatives 4.0 International licence (https://creativecommons.org/licenses/by-nc-nd/4.0/) which permits reproduction and distribution for non-commercial use without further permission provided the original work is attributed.

Open access book funded by Norwegian Institute of International Affairs.

British Library Cataloguing in Publication Data
A catalogue record for this book is available from the British Library

ISBN 978-1-5292-4892-0 paperback
ISBN 978-1-5292-4893-7 ePub
ISBN 978-1-5292-4894-4 OA ePdf

The right of Elana Wilson Rowe, Paul Beaumont and Lucas de Oliveira Paes to be identified as editors of this work has been asserted by them in accordance with the Copyright, Designs and Patents Act 1988.

All rights reserved: no part of this publication may be reproduced, stored in a retrieval system, or transmitted in any form or by any means, electronic, mechanical, photocopying, recording, or otherwise without the prior permission of Bristol University Press.

Every reasonable effort has been made to obtain permission to reproduce copyrighted material. If, however, anyone knows of an oversight, please contact the publisher.

The statements and opinions contained within this publication are solely those of the editors and contributors and not of the University of Bristol or Bristol University Press. The University of Bristol and Bristol University Press disclaim responsibility for any injury to persons or property resulting from any material published in this publication.

Bristol University Press works to counter discrimination on grounds of gender, race, disability, age and sexuality.

Cover design: Andrew Corbett
Front cover image: Stocksy/Oxana Pervomay

Contents

List of Figures and Tables   iv
Notes on Contributors   v
Acknowledgements   viii

1. Nature and Order in World Politics   1
   Lucas de Oliveira Paes, Paul Beaumont and Elana Wilson Rowe
2. Governing Peace and Security in the Anthropocene   20
   Dahlia Simangan
3. The Anthropocene, Climate Change and (Ecological) Security   42
   Matt McDonald
4. Nature's Hierarchies? Ecosystems and Order Making   62
   Elana Wilson Rowe, Paul Beaumont and Lucas de Oliveira Paes
5. To Unveil Nature's Secrets: International Cooperation in the International Geophysical Year   88
   Joanne Yao
6. Outer Space and Sovereignty in Post-Planetary Politics   112
   Katharina Glaab
7. World (Re)Ordering through Green Growth and Degrowth Futures   127
   Bruna Bosi-Moreira and Matthias Kranke
8. Seeing Like a Planet: Conclusion and Pathways for International Relations Scholarship   143
   Paul Beaumont, Lucas de Oliveira Paes and Elana Wilson Rowe

Index   155

List of Figures and Tables

Figures

4.1	Map of protected areas of the Caspian Sea showing river tributaries and nearby countries	71
4.2	Map of plastic input into the Arctic Ocean	72
4.3	Amazon ecosystem boundaries and forest carbon density	77
5.1	International Geophysical Year three-cent 1958 issue US stamp	100
7.1	Environmental futures and (dis)order	129

Tables

2.1	Three premises for rethinking and governing peace and security in the Anthropocene	22
7.1	Environmental futures and IR concepts	137

Notes on Contributors

Paul Beaumont is a senior researcher at the Norwegian Institute of International Affairs. His research interests include international relations theory, the (dis)functioning of international institutions, global environmental politics, nuclear weapons and hierarchies in world politics. He has published two monographs: *Performing Nuclear Weapons: How Britain Made Its Bomb Make Sense* (2021) and *The Grammar of Status Competition: International Hierarchies and Domestic Politics* (2024). His research has also featured in several IR journals, including *European Journal of International Relations*, *International Studies Review* and *Contemporary Security Policy*. He is currently an editor of the journal *Cooperation and Conflict*.

Bruna Bosi-Moreira is a postdoctoral researcher (Wissenschaftliche Mitarbeiterin) at the Institute of Political Science of the Friedrich-Alexander-Universität Erlangen-Nürnberg. She earned her PhD in political science and IR at both the University of Duisburg-Essen and the University of Brasilia. Her research interests lie at the intersection of the climate–energy nexus and geopolitics.

Katharina Glaab is Associate Professor at the Department of International Environment and Development Studies at the Norwegian University of Life Sciences. Her research focuses on IR theories, norms and knowledge practices in global politics, and the politics of environmental and sociotechnical changes in the Anthropocene. She is leader of the 'Nordic Space Infrastructures' research project funded by the Norwegian Research Council.

Matthias Kranke is Junior Professor of Global Sustainability Governance at the College for Social Sciences and Humanities of the University Alliance Ruhr, Faculty of Social Sciences, University of Duisburg-Essen, Germany. His research focuses on the global governance of intersecting social, economic and ecological issues, especially the tensions between the norm of economic growth and the goal of socioecological sustainability.

His work has been published in various journals, including *European Journal of International Relations*, *Global Policy*, *Global Society*, *Global Studies Quarterly*, *Review of International Political Economy* and *Review of International Studies*.

Matt McDonald is Professor of International Relations in the School of Political Science and International Studies at the University of Queensland. His research examines security, global climate politics and their relationship. He has published widely on these themes and is the author of *Ecological Security* (2021) and *The Environment, Security and Emancipation* (2012), and is co-author of *Ethics and Global Security* (2014).

Lucas de Oliveira Paes is a senior researcher at the Norwegian Institute of International Affairs. His research focuses on hierarchical dynamics and asymmetries in exercising state sovereignty across multiple international order issues. His main focus is power dynamics in global environmental governance, having conducted postdoctoral research on Amazonian regionalism as part of a comparative project on transboundary ecosystem governance funded by the European Research Council, and he is now leading a four-year project on multilevel governance networks in the Amazon and the Gulf of Guinea funded by the Research Council of Norway. His research has been recently published in journals such as *Environmental Politics*, *Cooperation and Conflict*, *Marine Policy*, *Review of International Studies*, *International Affairs* and *Journal of Peace Research*.

Dahlia Simangan is Associate Professor of International Relations at Hiroshima University. Her research interests in peace and conflict include topics on postconflict peacebuilding, the relationship between peace and sustainability, and international peace and security in the Anthropocene. She is the author of *International Peacebuilding and Local Involvement: A Liberal Renaissance?* (2019) and has written articles in leading IR and peace studies journals. She is Associate Editor of the *Review of International Studies*, Assistant Editor of *Peacebuilding* and a member of the Planet Politics Institute.

Elana Wilson Rowe is Professor of Global Governance at the Norwegian University of the Life Sciences and an associated research professor at the Norwegian Institute of International Affairs. Her research areas include the role of environmental governance in global order and polar and ocean governance. She is the author of *Russian Climate Politics: When Science Meets Policy* (2013) and *Arctic Governance: Power in Cross-Border Relations* (2018). She led a five-year research project funded by the European Research Council, comparing the politics of the Arctic, the Amazon basin and the Caspian Sea.

Joanne Yao is Reader in International Relations at Queen Mary University of London. Her research interests are in IR theory, environmental history and politics, global historical sociology, critical geopolitics, the history of empire and imperialism, and international organizations and institutions. Her first book, *The Ideal River* (2022), examines the construction of the 'ideal river' in the European geographical imagination and the establishment of the first international organizations. It won the 2023 LHM Ling Outstanding First Book Prize from the British International Studies Association and the 2024 Harold and Margaret Sprout Award from the International Studies Association.

Acknowledgements

The editors want to thank, first of all, the contributors to this volume for believing in the project, producing groundbreaking work from their fields of expertise, and thoroughly and thoughtfully considering how their insights on the governance of nature can help rethink IR. We also want to acknowledge the broader team of the European Research Council (ERC)-funded LORAX project (Horizon 2020 #803335, *The Lorax Project: Understanding Ecosystemic Politics*). Our Lorax project colleagues Kristin Fjæstad and Cristiana Maglia commented on numerous early drafts of the concept documents for this book and have been central to the broader intellectual project underpinning the volume.

We benefited from input of the scholarly community at two events: the 'Global Environmental Politics Reversed' roundtable at the International Studies Association annual conference in 2024 and our panel at the European International Studies Association annual conference 2024. Here, Paula Haufe deserves a special mention for her insightful comments as stand-in discussant in Lille.

There are many different types of support that lead to new scholarship. We extend our gratitude to the science committees and project administrators at the ERC who have been supportive and helpful, demonstrating that the pursuit of innovative research is a guiding principle for the ERC. At NUPI, the Lorax project as a whole and this book in particular would not have been possible without the support of our research administration colleagues Maryam Sugaipova, Jonny Andreassen and Kjersti Rånes Haugan.

Our editor at Bristol University Press, Zoe Forbes, deserves enormous credit for her astute guidance through the acquisition, peer-review and publishing processes. We have benefited from and are grateful for the professionalism and quality of the Bristol University Press team overall. Last but not least, we are also incredibly grateful for the anonymous reviewers, whose close and constructive feedback across multiple rounds helped improve the manuscript through to its final published version. We are especially thankful to one reviewer who saw things through from start to finish and contributed to improving the book by degrees with consistently sharp and generous critiques and questions.

1

Nature and Order in World Politics

Lucas de Oliveira Paes, Paul Beaumont and Elana Wilson Rowe

Introduction

The Pikialasorsuaq as it is known in Inuktitut and Greenlandic – or North Water Polynya in English – is an area of year-round open water surrounded by sea ice only slightly smaller than the UK. This polynya teems with life and is a breeding ground and migration area for animals such as narwhal, beluga, walrus, bowhead whales and migratory birds. While the area now stands divided between two modern polities – Greenland and Canada – the polynya is comprised of the traditional hunting grounds and travel routes of Inuit living on both sides. In 2017, the Pikialasorsuaq Commission was established as an Inuit management authority, led by local communities. Communities from both countries would consult and discuss potential questions of shipping, mining, tourism or other economic activities that might increase as the sea ice extent decreases due to the impacts of global climate change. Certainly, challenges abound – the interests of the 129 people in Aujuittuq on the western side of the polynya and the 646 individuals in Qaanaaq on its eastern side pale alongside those of the comparatively giant capital cities of Greenland, Nuuk and faraway Ottawa, Canada. The shores of the Pikialasorsuaq also house the Pituffik Space Base (formerly known as Thule Air Base) owned by the US, involving the security interests of well over 331 million Americans (and more if key allies are taken into consideration). Despite difficult implementation challenges to navigate, the Pikialasorsuaq Commission effort is remarkable and significant. It entails envisioning and pursuing a new kind of regional order in which Indigenous peoplehood is respected and an area that had become 'international' only through the practices of state colonialism is decolonized (Pikialasorsuaq Commission 2017; Wilson Rowe 2024).

The politics of the Pikilasorsuaq remind us that we need to be analytically alert to how the governance of nature and world ordering are interconnected. Climate change and the crises of nature are foregrounding global environmental challenges in politics, jumbling together previously separate fields of policy making from human rights through trade to security. Specific challenges that we now group together as linked global crises of climate and nature have long been addressed – if not necessarily successfully – through several global environmental governance negotiations and initiatives, such as those pertaining to biodiversity, climate mitigation and adaptation, treaties to minimize pollutants and so on. These efforts have inspired an expansive and important research programme that systematically explores when, why and how these global environmental governance efforts work to address the problems they are established to solve.

Yet there remains a pressing need to understand how the challenges of the Anthropocene are generating, reshaping and even transforming core practices of global politics more broadly, as well as our conceptual frameworks for understanding them. As a growing body of work points out, governance of nature has, does and can have broader consequences for power relations and world ordering (Corry 2013; Dalby 2014, 2020; McDonald 2013, 2021; Aykut, Foyer and Morena 2018; Corry and Stevenson 2018; Yao 2019, 2021, 2022; Wilson Rowe 2021; Aykut and Maertens 2022; Beaumont and Wilson Rowe 2022; Paes 2022, 2023; Maglia and Wilson Rowe 2023, 2024). In other words, there is also a politics *of* the environment that is not necessarily *for* the environment.

This volume aims to bring to the forefront how the governance of nature has consequences for how global politics is ordered more broadly. Instead of looking at the effects of the structures and process of the current international order over nature, we look at how the governance of nature also serves to constitute and shore up important aspects of world order. Indeed, each of our chapters blurs the boundaries between the enactment of environmental politics and how global politics is patterned, structured and understood more generally. More specifically, the contributions explore empirically and theoretically how recognition and governance of environmental issues (or perhaps ecological omni-crises) is transforming or has transformed global politics and governance writ large. In the following chapters, we consider this question against and through core concepts for scholarship in international relations (IR), from security through territoriality and hierarchy to cooperation. Answering these questions not only promises to shed new light on global governance practices but also prompts reflexivity within IR around the adequacy of its prevailing theoretical apparatus.

This volume is not intended to be a grand theoretical exercise supplanting diversity among critical environmental scholars in IR or problem-solving approaches to global environmental challenges. Instead, our goal is humbler: the volume aims to highlight commonalities among a growing strand of research in IR engaged with the historical, current and emerging links between the politics of nature and world order, a link that can no longer be overlooked amid the transformations of world politics in the Anthropocene. In so doing, we identify, bring into dialogue and amplify a hitherto diffuse collection of IR research. While perhaps not (yet) unified by a shared a research agenda, the chapters in this volume, taken together, identify and showcase two complementary angles for studying the governance of nature and the making of world order: (1) exploring empirically and then theoretically how the governance of nature has shaped international order; and (2) critically interrogating the utility (or lack thereof) of core concepts for IR scholarship and global governance practice in coming to terms with how the governance of nature has, does and will shape and constitute world order more broadly. We contend that taking stock of these strands of research renders visible the contours of a potential but overdue new 'great debate' in IR: one that would force the 'isms' into a reckoning with whether, how and to what extent the existing conceptual apparatus – within the policy world and academia – remains tenable amid the global omni-crises of the Anthropocene. As we will outline in the conclusion, such a debate need not imply supplanting theoretical or methodological diversity with a single 'winner', but encouraging the 'isms' of IR into more thoroughgoing dialogue with ongoing global environmental transformations and their political implications.

A green gap in international relations? Empirical and disciplinary challenges

When global environmental politics (GEP) emerged as a subfield within IR in the 1980s, it was something of a niche issue. The early international negotiations around environmental issues – such as atmospheric ozone depletion – were undeniably international, but also seemingly of only marginal significance to the so-called high-politics of interstate relations and to the discipline of IR itself (Litfin 1994). Yet, the specific conditions of this niche – the ostensibly low political stakes, the undeniable significance of (hard) science and the possibility of identifying clear win-win scenarios – provided the empirical conditions for (once maligned) liberal theories of IR to thrive. Robert Keohane, among many others, was only too keen to point out the obvious: realisms – both neo and classical – offered little purchase in terms of understanding the international institutional arrangements that had mushroomed throughout the 1980s and 1990s (Keohane 1986). The liberal perspective invigorated the inquiries of scholars working to make sense of burgeoning environmental governance as they developed a conceptual apparatus to analyse these phenomena and provide the evidence basis for improving their design and functioning (for example, Haas 1989; Young 1989; Bernauer 1995; Martin and Simmons 1998).

Indeed, mainstays of IR today – regime theory, neoliberal/rational institutionalism and so on – were cultivated within GEP and arguably depended upon the prevalence of empirically observable cases of successful international cooperation (O'Neil 2009: 13; Corry and Stevenson 2017: 12–13). Here, science and technology identify problems and potential solutions that rational interest-driven states could cooperate in order to implement. Crucial to this analytical agenda was the normative aspiration that international institutional design could serve as an intervening variable that could share information, establish expectations, address distributional concerns and thereby facilitate long-term win-win arrangements. Having ostensibly shown how many states could cooperate effectively in environmental matters, neoliberal institutionalism sought to explore whether and when they would or could in higher politics too (Martins and Simmons 1998; Wallendar 2000). The success in explaining and informing institutionalized cooperation post-Cold War era would prompt one leading IR scholar to declare liberalism the new dominant approach of both theory and practice (Sterling-Folker 2015: 45). While the rise to dominance of liberal theorizing has several causes (Sterling-Folker 2015: 45), the study of global environmental politics certainly helped establish liberal-institutionalist dominance in IR thinking, which in turn structured the evolution of GEP as a subfield.

This liberalism-informed GEP proceeded to develop into a thriving subfield of IR, producing research to help understand and address a set of previously overlooked issues relating to the governance of nature. GEP scholarship has generated research that helps us make sense of the complex suit of complex of regimes and institutional arrangements devised to govern our planet (Raustiala and Victor 2004; Andonova, Betsill and Bulkeley 2009; Keohane and Victor 2011; Green 2013; Abbott, Green and Keohane 2016) and discerning their relative effectiveness (or lack thereof) (Dimitrov 2005, 2020; Chan et al 2015; Jordan et al 2015; Green 2021; Biermann et al 2022). GEP has also shed light on the complex constellation of political interests driving environmental problems and illuminating institutional instruments to address them (Paterson and Grubb 1992; Newell and Paterson 2010; Falkner 2016; Hale 2020; Colgan, Green and Hale 2021). This scholarship has contributed to understanding interplay across strands of global environmental governance, from structural arrangements, such as regimes, to novel relational mechanisms, such as orchestration (Abbott and Bernstein 2015). Similarly, the Earth System Governance project has brought into focus the broad structural patterns emerging from the institutional complexity of global environmental governance and on the effectiveness of such architecture (Biermann et al 2010; Burch et al 2019; Biermann and Kim 2020; Lövbrand, Mobjörk and Söder 2020).

This flourishing is well reflected in the number of GEP articles in leading IR journals, in the rise in the impact factor of GEP journals and in the general recognition among IR academics (not only within GEP) regarding the significance of environmental issues (see Green and Hale 2017). Hence, as Corry and Stevensen (2017: 197) discuss, IR has rediscovered its object of analysis' entwinement with the natural world through engagement with environmental 'issues'. Yet, the challenge that two longstanding scholars of environmental politics identified almost a decade ago remains: the focus on environmental issues as something somehow 'apart' is highly problematic given that 'environmental problem-solving is no longer concerned with isolated problems, but rather with reorganizing the overall relation between humans and natural systems' (Pattberg and Widerberg 2015: 684). Pattberg and Wideberg continue by stating that 'while technical debates about the "right" governance instruments to combat climate change are intellectually defensible, successful climate governance will crucially depend on our ability to understand the interactions of politics, economics, ideology and the human condition. In short, our current cross-cutting "deep" problems require holistic theorizing' (2015: 704). While this trend may well (hopefully) be reversing, it points to a dominant view in IR in which environmental problems are seen as compartmentalized, a disciplinary gap to be filled through studies of environmental politics.

We believe that the compartmentalization of GEP as a separate subfield is tied to the very notion of a 'green gap' in IR that is waiting to be closed while broader political processes (and the rest of IR) carry on as before. Indeed, the designation of environmental issues to a separate subfield thus functions to secure 'old IR' from geophysical and associated political processes that it has always been embedded within and co-productive of, but historically preferred to 'airbrush' out (Corry and Stevenson 2017). On this point, Corry and Stevenson warn against simply adding 'environment to existing theoretical frameworks for understanding global politics – and conversely adding "governance" to natural science analysis of Earth Systems'. While they suggest that this approach 'has got us some of the way' to understanding how IR and the earth's biogeophysical systems concatenate, 'it is clearly not sufficient' (Corry and Stevenson 2017: 194). To be sure, the problem-solving focus of mainstream GEP is all too warranted and crucial to helping international society address its impacts on the natural world. However, such a focus may lose sight of the multiple ways in which the practices in the governance of nature has been constitutive of the several aspects of world order which are currently seen as hurdles for environmental governance per se. The broader constitutive relationship between practices of governance of nature and broader practices of world ordering is the focus of this volume.

Nature and ordering amid planetary crises

The pressures on planetary nature and climate are likely to render increasingly evident the constitutive role of the environment to global order. Numerous scientific assessments and an ever-growing number of coalitions have pointed to the consequences of humanity reaching a 'planetary boundary', beyond which human impacts driving global-level change in climate can no longer be reversed or slowed down (Rockström et al 2009, 2023). Reaching such a planetary boundary implies fundamental changes to the physical conditions against which our current societies have developed – these so-called 'Holocene conditions' will be irretrievable. Changes are ongoing via gradual accretion and the prospect of some critical ecosystem 'tipping points' that could precipitate dramatic planetary change has gained increased scientific and policy attention (Lovejoy and Nobre 2018, 2019; Lenton et al 2019; Pereira and Viola 2020; Gatti et al 2021; Milkoreit et al 2024).

The prospects and ongoing manifestations of these enduring changes result in a widespread perception that we are witnessing an omni-crisis heralding multiple and unpredictable tectonic shifts. This emerging period in which we have entered is often termed the 'Anthropocene epoch', in which the impacts of humans and their activities create a 'rupture in both human and Earth history' and usher in a period in which natural and human forces are so intertwined that change in one will trigger change in the other (Lundborg 2023, 599). While the urgency and dimension of such crises make the need for scholarship focused on tackling environmental problems ever more important, the sheer scale of the task before us invites us to question its deeper roots and broad ramifications for the making of the world order producing it. In this sense, the need for rethinking disciplinary divisions and subdivisions and to engage in holistic theorizing is becoming ever more pressing as global climate and natural change associated with the Anthropocene has begun to weigh in on previously disparate fields of governance (for example, Oels 2012; Burke et al 2016; Fishel et al 2018).

Along these lines, several strands of critical environmental scholarship have already started to disrupt disciplinary boundaries in order to delve into the broader transformations taking place in the Anthropocene. A burgeoning body of work provides *both* a critique of prevailing anthropocentrism within global environmental governance and seeks to develop an alternative normative framework and agenda (Corry 2013; McDonald 2013, 2021; Dalby 2014). Part of this scholarship proposes a shift of analytical focus away from actors and their relations to one another and towards the process of constructing our planet as a set of governance *objects* that pull multiple actors and power relations around them (Corry 2013, 2024; Esguerra 2024). Other works go further to problematize the binaries – human/nature, political/nonpolitical and so on – that underpin humanity's approach to and relationship with nature and bring forward alternative, marginalized and/or previously forgotten approaches to the human–nature relationship. There are numerous strands to this work – posthumanism, Indigenous/postcolonial, deep ecology – that are conducting the necessary intellectual labour to imagine and thereby expand global imaginaries in light of the Anthropocene (Chandler 2018; Inoue 2018; Castro Pereira and Saramago 2020; Brigg, Graham and Weber 2022; Reddekop 2022). This critical scholarship in IR thrives through and in close dialogue with the wider social sciences and humanities, but often concentrates conversations in the field's own journals and conferences. Like GEP in general, this critical strand has flourished under the 'theoretical peace' that followed IRs third great debate, which allowed a diversity of approaches to IR to thrive at the expense of discipline-wide common conversation on core theories and methods (Sylvester 2013; Heiskanen and Beaumont 2024).

This volume also takes inspiration from critical scholarship that has rightfully warned that how we define or even name our planetary crisis has consequences for how we make sense of the social structures and processes that have produced it, as well as for understanding the roles of different social groups in causing and addressing such crises. Indeed, the term 'Anthropocene', with its implication of a monolithic 'humanity' as a planetary force, risks eliding important differences in contribution and vulnerability to ecological crisis (Diffenbaugh and Burke 2019; Simangan 2020; Chipato and Chandler 2023a). Encapsulated in the slogan 'system change, not climate change', these works highlight how crises of climate and nature that fall under the term 'Anthropocene' are inseparable from the multiple longstanding global inequalities.

Taking up this demand for system change rather than climate change, this scholarship generally highlights the multiple binaries and ideologies that must be considered in light of Anthropocene change. Of central importance for interrogating the prospects of system change is the selection of which binaries become core points of analysis. While it may seem obvious to put human–nature relations at the heart of Anthropocene inquiry, such a focus can gloss over or even reproduce existing binaries of modernism, and the unjust racialized and colonial hierarchies and structures that such binaries support (Chipato and Chandler 2023a, 2023b). These scholars draw upon insights from Black studies and Black feminism to argue that:

> the most important ontological divide is the one which is constitutive of modernity, that between those who have the capacity for 'being' (the human) and those who lack 'being' (the black(ened) non or

semi-human). It is this divide which allows for the exploitation of those considered Black as objects and the instrumentalisation of the objects that are created through this divide. (Chipato and Chandler 2023b: 279)

In other words, which binaries and whose Anthropocene crisis are essential questions to be posed. While this volume considers the Anthropocene to provide a useful heuristic for capturing the impacts and sense of planetary crisis shaping world politics today, we also consider it crucial to be sensitive to its analytical and ethical elisions, as the chapters by Simangan, Bosi-Moreira and Kranke and McDonald elaborate.

The seeming mismatch between the technocratic problem solving of liberal institutionalism and the scale and complexity of earth's ecological omni-crises is likely a big part of the reason why a burgeoning body of critical scholarship has called into question the adequacy IR's frameworks and the associated epistemologies, ontologies and methodologies (McDonald 2013, 2021; Pereira and Saramago 2020; Simangan 2020). Indeed, some explicitly seek to 'end IR' as we know it and instead endorse a manifesto for 'Planet Politics' (Burke et al 2016). From this perspective, global environmental politics has always been global politics and global politics has always been global environmental politics, but it has only recently become untenable to pretend otherwise and imperative to recognize it.

Building upon these efforts to bring the conceptual apparatus of IR up to speed, this volume casts the so-called green gap in IR in a new light. The chapters illuminate environmental governance's entanglement in the transformation of world ordering writ large and highlight growing scholarship that addresses these deep interlinks between the structures, institutions and processes devised to govern the natural world and structures, institutions and processes devised to govern other dynamics of world politics. Hence, the chapters also demonstrate how the leading edge of IR lends itself well to exploring how the governance of nature has shaped and is shaping order, as well as critiquing the conventional wisdom within global governance and IR. While little of this research has yet to make it into the core textbooks of IR teaching, it is certainly a growing and fruitful avenue of inquiry. In other words, we may (or should) soon cease talking about closing the green gap and rather about *further* problematizing the boundaries between the environmental and 'non-environmental issues'.

To this end, this volume aims to foster dialogue, amplify and encourage future research traversing the nature/governance intersection. This is important not only for disciplinary development but also because IR as a discipline historically has a close conversation with policy and policy making. Categories of political practice and scholarly analysis have informed one another in foreign policy, international institutions, security and so on, traversing university lecture theatres and ministry meeting rooms – albeit often with a significant time lag and unintended consequences (Musgrave 2021; Hendrix et al 2023; Lerner and O'Loughlin 2023).

The contributions in this volume help us to rethink world politics amid our multiple planetary crises and consider the intertwinement of the governance of nature and world order. We do so by looking at the ways in which dynamics associated with governing nature have shaped or are transforming key processes, institutions, ideas and concepts usually deemed constitutive of the world order. Our focus on order stems from the view that it is a concept that encapsulates multiple political projects that have constituted our world as it is. To be sure, the notion of order is a rather elusive one, around which much theoretical attention has been devoted. However, we believe that such elusiveness invites theoretical dialogue and pluralism in search of making sense of the key processes, practices and relations that shape world politics. The theoretical diversity underlying the concept of order is evidenced in how most works fall back to the all-embracing definition of order as 'settled rules and arrangements between states that define and guide their interaction' (Ikenberry 2012: 12) or, even more broadly, 'patterns of activity' that structure and orient world politics (Bull 1977). Hence, part of the appeal of the study of order is that it indeed encompasses a broad array of structures and practices, while inviting common reasoning about their constitutive role in shaping the usually taken-for-granted defining features of world politics. Analytically, we contend that investigating world order can be fruitfully undertaken by revisiting the key concepts around which the study of IR has evolved. Therefore, understanding the entanglement of nature's governance and global ordering requires understanding how nature has shaped or is transforming dynamics associated with such concepts.

Core questions and structure of the book

In this volume, we seek to highlight the disciplinary and policy gains that can be made when we consider how the governance of nature (including environmental politics) shapes world order more broadly. In this sense, we pick up on the challenge outlined previously about overcoming the 'green gap' by not treating it as a niche to be filled, but rather by revisiting key IR scholarly concepts to explore how far they take us in terms of understanding the interplay between ordering nature and ordering world politics. In the following chapters, we argue that the governance of nature needs to be understood as fundamental to the practices of world politics – and thus to IR's conceptual frameworks. The chapters consider in turn core ideas that structure relations in global politics and our scholarly understanding of them and illustrate how the governance of nature shapes, challenges and/or transforms these core notions.

The chapters in this volume engage with the following questions, bringing to the fore several key cases of how governing nature reshapes the practices of global politics, and our conceptual apparatus for understanding IR:

- How does the governance of nature shape practices of world ordering from the pursuit of sovereignty through security and peace to multilateral coordination and cooperation?
- How do key concepts of IR and world politics enable and limit our understandings and the processes of governing nature and making order?
- How will the governance of nature, including responding to the omni-crises associated with the Anthropocene, shape world order and IR as a discipline?

The opening two chapters take a panoramic view of IR the discipline and field of practice and make the case that the omni-crisis of the Anthropocene requires a conceptual revolution. In Chapter 2, Simangan invites us to rethink the broader architecture of multilateral governance peace and security on novel grounds that recognize the pressing and complex reality of the Anthropocene, where nature and culture blend and political boundaries blur. She argues that the Anthropocene disrupts the binaries around nature and culture, while highlighting the porousness of territoriality and the conflation of temporalities. For these reasons, the Anthropocene provides the philosophical landscape for rethinking peace and security in accordance with environmental considerations. The chapter thus argues that it is imperative to rethink vulnerability, militarism and growth as they entrench the anthropocentric, state-centric and linear approaches to governing peace and security. Ultimately, it argues that a radical change of the foundations of the current world order is not only desirable within the current logics of peace and security, but is also necessary for humanity's survival in the Anthropocene.

In Chapter 3, McDonald makes the analytical and ethical case for governing security on ecological terms, recognizing the systemic nature of Anthropocene threats. He counters the traditional conceptualization of security in IR which suggests self-contained units attempting as best they can, at times in cooperation with each other, to protect themselves from external threats to the integrity of those units. Whereas this conception is central to prevailing notions of national security and international security, the Anthropocene context and the scale of the contemporary ecological crises expose fundamental limits to this way of conceiving and approaching security. In relation to climate change, the conventional conception of security encourages states to insulate themselves from the effects of a changing climate and work through international organizations to attempt to minimize spillover effects from those that are unable to do so. McDonald argues that both dynamics are happening already, but the global nature of climate change points to the inability of even powerful states to wholly protect citizens from its effects, even with unprecedented unilateral action. At the same time, the ongoing generation of the problem through everyday activity, production and economic exchange suggests – for powerful states in particular – that the threat is not external, but rather is internally generated. Therefore, the chapter points to the limited and limiting pathologies of traditional approaches to security in IR when coming to terms with ecological crises, before making a case for ecological security as a means of guiding effective and ethically defensible responses to those crises.

The next two chapters then explore how rule over nature has shaped key features and our contemporary order. In Chapter 4, Wilson Rowe, Beaumont and de Oliveira Paes analyse the broader effects of growing efforts to govern the world's ecosystems. Such ecosystems do not respect sovereign borders; hundreds traverse more than three states and thus require complex international cooperation. The chapter builds upon an emerging research agenda that critically examines the political and social consequences of these cooperative arrangements anchored in transboundary ecosystems. Synthesizing multiple case studies that have explored the effects of cooperation around the Arctic, Amazon and Caspian Sea, the chapter shows how these very different cases of ecosystem politics generate similar consequences. Indeed, examining ecosystem politics through a hierarchy lens, it shows how the process of scaling and spatializing along the contours of the ecosystem generates three types of interrelated hierarchical dynamic that would otherwise prove unlikely to obtain were cooperation organized through other means. First, all three cases highlight how ecosystems bring together unlikely bedfellows into a new political grouping and collective identity. In this sense, ecosystem politics overcome or become insulated from prior or exogenous geopolitical tensions. Second and relatedly, the closer insider relations beget and legitimate sharper boundaries demarking insiders from outsiders (non-ecosystem adjacent states) and facilitate regional exclusionary dynamics. As the cases show, these can and do enable the inversion of global hierarchies, regional gatekeeping and the empowerment of conventionally weaker actors. Finally, the most advanced form of ecosystem politics involves the thickening of social relations within the group to the point that the ecosystem insiders can act as coalition in global politics, beyond the region or policy field within which it was originally established. Taken together, the chapter argues that, with the acceleration of the Anthropocene, governing from (if not for) ecosystems could eventually shape and even transform the international order.

In Chapter 5, Yao unpacks how the endeavour of constructing the planet as an object of governance around the International Geophysical Year (IGY) of 1957–1958 was crucial to shield international scientific cooperation from Cold War rivalry, but also entailed the reproduction and elision of colonial hierarchies. In a number of arenas of global environmental governance, science is celebrated as a unifying force – from the Antarctic Treaty System (ATS) that governs the southern continent, to the United Nations Framework Convention on Climate Change (UNFCCC), to the international space station. In these forums, science serves as a legitimating common goal and collaborative practice, which promises to help international society rise above the politics that hinder cooperation around the governance of nature. Problematizing this narrative, the chapter examines the proposition that science offers an escape from politics by interrogating the IGY 1957–1958, which led to the 1959 Antarctic Treaty. First, the chapter explores how scientists and diplomats engaged in the IGY and the ATS framed science as a solution to the political impasses of the Cold War. It then critically examines to what extent the projects were actually apolitical. It argues that, rather than an escape from politics, these scientific projects to complete human knowledge of the global environment were deeply political and shaped by longstanding global hierarchies from the age of empire. It concludes with a reflection of how scientific projects to increase our knowledge of nature are political projects and efforts to frame them as an escape from politics reinforce the hidden hierarchies that lurk beneath their technocratic surface.

The final two empirical chapters consider future transformations stemming from the current dynamics of nature's governance. In Chapter 6, Glaab looks at the governance of nature beyond the Earth, examining whether and how it may challenge prevailing notions of territorial sovereignty. Outer space usually does not feature in considerations of environmental politics; planetary politics stops with the invisible atmospheric border. This is curious as not only have satellite images of Earth from outer space enabled the visualization and construction of the Anthropocene, but nature clearly also extends beyond the planetary realm and can be considered an additional ecosystem. In light of a new space race which expands economic activities to outer space to extract resources, develop industries or space tourism, while at the same time building infrastructures and economies on Earth, questions around the sustainability of space activities come to the fore, leading to calls for 'space environmentalism' and new regulatory initiatives. But what does it mean when efforts to govern nature extend beyond planetary boundaries and include outer space and what kind of implications does it have for *global* governance? The chapter first critically discusses the reflex of IR scholarship to explain conflicts as a matter of geopolitics. In this reading, outer space is usually conceptualized in IR as an extension of power rivalries on Earth, where nature often only features as a resource that can be extracted and acquired, or territory that can be gained. In a second step, the chapter shows how an understanding of outer space as a global commons problem opens the door for liberal-institutionalist perspectives on collective action. While this perspective accounts for environmental issues in outer space, this chapter goes one step further: it shows how a relational ontology helps not only to conceptualize Earth and space as co-constitutive, but also disrupts binaries of not only nature/human but also earth/space and human/nonhuman. This highlights that governance aiming at regulating environmental issues related to space activities cannot be separated from the geopolitics of outer space.

In Chapter 7, the final empirical chapter, Bosi-Moreira and Kranke foreground the ordering effects of environmental governance by critically examining 'green growth' and 'degrowth' visions for the future of our planet. The chapter's fundamental premise is that, as the construction of nature as a governable object matters for processes of world ordering, the imaginaries of the future deserve particular attention because of their potential to influence politics in the present. The chapter discusses how the literature on environmental futures has been spread across fields, such as anthropology or science and technology studies, but is still to be consolidated in IR. By connecting research on environmental futures with IR scholarship on order, the chapter shows how envisioning a greener economy can influence world ordering and the governance of nature. The authors develop this line of reasoning with respect to the two future-oriented approaches to global environmental governance – 'green growth' and 'de-growth'– each of which envisions very different future world orders and brings about very different implications for the set of social transformations needed to achieve that future.

Conclusion

This book examines how governance of nature constrains, supports or constitutes world ordering processes and considers the extent to which IR's conceptual apparatus is equipped to make sense of such dynamics. Each chapter problematizes the nature-order nexus by revisiting key concepts of IR and exploring different time periods, thereby ensuring that this book reflects on historical as well as more current cases of how the governance of nature shapes world order. Hence, the volume not only demonstrates that the governance of nature has been and is constitutive of world order, but that it also does so in multiple interlinked ways. In so doing, the volume aims to catalyse an already emerging agenda for studying the multiple ways in which the governing of nature and the making of world order are imbricated.

Taken together, the chapters in this volume highlight how ongoing work in disparate strands of IR are engaging with the link between nature and order, thereby contributing to an enhanced understanding of how the governance of nature (historically and in today's ecological crisis) leads to intended and unintended transformations that must be better understood. Ultimately, the volume acts as a collective warning against the longstanding tendency in IR theorizing to bracket nature and the environment as self-contained issues, apart from broader ordering dynamics. Thus, as well as advancing our understanding of the links between nature's governance and world order, we contend that this multi-pronged collective effort ultimately poses major questions pertaining to the adequacy and the efficacy of IR's key concepts and thus the ontological assumptions about what international politics *is* and what doing IR should *entail*. As we argue in the conclusion, the chapters suggest that the study of nature's governance as a field should not be limited to the purview of global environmental politics as a subfield of IR. Instead, they show the necessity of theorizing IR as embedded in practices of nature's governance – that is, inscribed *within* global environmental politics. To ferment such a reckoning within the discipline and its silos, we conclude this book by arguing that the time is nigh for IR's 4th great *(environmental)* debate between the growing numbers calling for a conceptual revolution and those that appear to believe IR can carry on as before.

References

Abbott, Kenneth W. and Steven Bernstein. 2015. 'The High-Level Political Forum on Sustainable Development: Orchestration by Default and Design'. *Global Policy* 6(3): 222–233.

Abbott, Kenneth W., Jessica F. Green and Robert O. Keohane. 2016. 'Organizational Ecology and Institutional Change in Global Governance'. *International Organization* 70(2): 247–777.

Andonova, Liliana B., Michele M. Betsill and Harriet Bulkeley. 2009. 'Transnational Climate Governance'. *Global Environmental Politics* 9(2): 52–73.

Aykut, Stefan Cihan, and Lucile Maertens. 2022. 'The Climatization of Global Politics: Introduction to the Special Issue', in *The Climatization of Global Politics*. Cham: Springer International Publishing, pp 1–18.

Aykut, Stefan Cihan, Jean Foyer and Édouard Morena (eds). 2018. *Globalising the Climate: COP21 and the Climatisation of Global Debates*. New York: Routledge.

Beaumont, Paul and Elana Wilson Rowe. 2022. 'Space, Nature and Hierarchy: The Ecosystemic Politics of the Caspian Sea'. *European Journal of International Relations* 29(2).

Bernauer, Thomas. 1995. 'The Effect of International Environmental Institutions: How We Might Learn More'. *International Organization* 49(2).

Biermann, Frank and Rakhyun Kim. 2020. *Architectures of Earth System Governance Institutional Complexity and Structural Transformation*. Cambridge: Cambridge University Press.

Biermann, Frank et al. 2010. 'Earth System Governance: A Research Framework'. *International Environmental Agreements: Politics, Law and Economics* 10(4): 277–298.

Biermann, Frank et al. 2022. 'Scientific Evidence on the Political Impact of the Sustainable Development Goals'. *Nature Sustainability* 5(9): 795–800.

Brigg, Morgan, Mary Graham and Martin Weber. 2022. 'Relational Indigenous Systems: Aboriginal Australian Political Ordering and Reconfiguring IR'. *Review of International Studies* 48(5): 891–909.

Bull, Hedley. 1977. *The Anarchical Society*. Cham: Springer.

Burch, Sarah et al. 2019. 'New Directions in Earth System Governance Research'. *Earth System Governance* 1: 100006.

Burke, Anthony, Stefanie Fishel, Audra Mitchell, Simon Dalby and Daniel J. Levine. 2016. 'Planet Politics: A Manifesto from the End of IR'. *Millennium: Journal of International Studies* 44(3): 499–523.

Castro Pereira, Joana and André Saramago (eds). 2020. *Non-human Nature in World Politics: Theory and Practice*. Cham: Springer.

Chan, Sander et al. 2015. 'Reinvigorating International Climate Policy: A Comprehensive Framework for Effective Nonstate Action'. *Global Policy* 6(4): 466–473.

Chandler, David. 2018. *Ontopolitics in the Anthropocene: An Introduction to Mapping, Sensing and Hacking*. New York: Routledge.

Chipato, Farai and David Chandler. 2023a. 'After the End of the World? Rethinking Temporalities of Critique and Affirmation in the Anthropocene'. *International Relations*. https://doi.org/10.1177/00471178231194710

Chipato, Farai and David Chandler. 2023b. 'Critique and the Black Horizon: Questioning the Move "beyond" the Human/Nature Divide in International Relations'. *Cambridge Review of International Affairs* 37(3): 1–19.

Colgan, Jeff D., Jessica F. Green and Thomas N. Hale. 2021. 'Asset Revaluation and the Existential Politics of Climate Change'. *International Organization* 75(2): 586–610.

Corry, Olaf. 2013. *Constructing a Global Polity*. Basingstoke: Palgrave Macmillan.

Corry, Olaf. 2024. 'Making the Climate Malleable? "Weak" and "Strong" Governance Objects and the Transformation of International Climate Politics'. *Global Studies Quarterly* 4(3). https://doi.org/10.1093/isagsq/ksae062

Corry, Olaf and Hayley Stevenson (eds). 2017. *Traditions and Trends in Global Environmental Politics: International Relations and the Earth*. New York: Routledge.

Dalby, Simon. 2014. 'Environmental Geopolitics in the Twenty-First Century'. *Alternatives* 39(1): 3–16.

Dalby, Simon. 2020. *Anthropocene Geopolitics: Globalization, Security, Sustainability*. Ottawa: University of Ottawa Press.

Diffenbaugh, Noah S. and Marshall Burke. 2019. 'Global Warming Has Increased Global Economic Inequality'. *Proceedings of the National Academy of Sciences* 116(20): 9808–9813.

Dimitrov, Radoslav S. 2005. 'Hostage to Norms: States, Institutions and Global Forest Politics'. *Global Environmental Politics* 5(4): 1–24.

Dimitrov, Radoslav S. 2020. 'Empty Institutions in Global Environmental Politics'. *International Studies Review* 22(3): 626–650.

Esguerra, Alejandro. 2024. 'Objects of Expertise. The Socio-material Politics of Expert Knowledge in Global Governance'. *Global Studies Quarterly* 4(3): ksae060.

Falkner, Robert. 2016. 'A Minilateral Solution for Global Climate Change? On Bargaining Efficiency, Club Benefits, and International Legitimacy'. *Perspectives on Politics* 14(1): 87–101.

Fishel, Stefanie, Anthony Burke, Audra Mitchell, Simon Dalby and Daniel Levine. 2018. 'Defending Planet Politics'. *Millennium: Journal of International Studies* 46 (2): 209–219.

Gatti, Luciana V. et al. 2021. 'Amazonia as a Carbon Source Linked to Deforestation and Climate Change'. *Nature* 595(7867): 388–393

Green, Jessica F. 2013. *Rethinking Private Authority*. Princeton: Princeton University Press.

Green, Jessica F. 2021. 'Does Carbon Pricing Reduce Emissions? A Review of Ex-post Analyses'. *Environmental Research Letters* 16(4): 043004.

Green, Jessica F. and Thomas N. Hale. 2017. 'Reversing the Marginalization of Global Environmental Politics in International Relations: An Opportunity for the Discipline'. *PS: Political Science & Politics* 50(2): 473–479.

Haas, Peter M. 1989. 'Do Regimes Matter? Epistemic Communities and Mediterranean Pollution Control'. *International Organization* 43(3): 377–403.

Hale, Thomas. 2020. 'Catalytic Cooperation'. *Global Environmental Politics* 20(4): 73–98.

Heiskanen, Jaakko and Paul Beaumont. 2024. 'Reflex to Turn: The Rise of Turn-Talk in International Relations'. *European Journal of International Relations* 30(1): 3–26.

Hendrix, Cullen S., Julia Macdonald, Ryan Powers, Susan Peterson and Michael J. Tierney. 2023. 'The Cult of the Relevant: International Relations Scholars and Policy Engagement beyond the Ivory Tower'. *Perspectives on Politics* 1–13.

Inoue, Cristina Yumie Aoki. 2018. 'Worlding the Study of Global Environmental Politics in the Anthropocene: Indigenous Voices from the Amazon'. *Global Environmental Politics* 18(4): 25–42.

Ikenberry, G. John. 2012. *Liberal Leviathan*. Princeton, NJ: Princeton University Press.

Jordan, Andrew J. et al. 2015. 'Emergence of Polycentric Climate Governance and Its Future Prospects'. *Nature Climate Change* 5(11): 977–982.

Keohane, Robert O. (ed.). 1986. *Neorealism and Its Critics*. New York: Columbia University Press.

Keohane, Robert O. and David G. Victor. 2011. 'The Regime Complex for Climate Change'. *Perspectives on Politics* 9(1): 7–23.

Lenton, Timothy M. et al. 2019. 'Climate Tipping Points: Too Risky to Bet Against'. *Nature* 575(7784): 592–595.

Lerner, Adam B. and Ben O'Loughlin. 2023. 'Strategic Ontologies: Narrative and Meso-level Theorizing in International Politics'. *International Studies Quarterly* 67(3). https://doi.org/10.1093/isq/sqad058

Litfin, Karen. 1994. *Ozone Discourses: Science and Politics in Global Environmental Cooperation*. New York: Columbia University Press.

Lövbrand, Eva, Malin Mobjörk and Rickard Söder. 2020. 'The Anthropocene and the Geo-Political Imagination: Re-writing Earth as Political Space'. *Earth System Governance* 4: 100051.

Lovejoy, Thomas E. and Carlos Nobre. 2018. 'Amazon Tipping Point'. *Science Advances* 4(2). https://doi.org/10.1126/sciadv.aat2340

Lovejoy, Thomas E. and Carlos Nobre. 2019. 'Amazon Tipping Point: Last Chance for Action'. *Science Advances* 5(12). https://doi.org/10.1126/sciadv.aba2949.

Lundborg, T. 2023. 'The Anthropocene Rupture in International Relations: Future Politics and International Life. *Review of International Studies* 49(4): 597–614.

Maglia, Cristiana and Elana Wilson Rowe. 2023. 'Ecosystems and Ordering: Exploring the Extent and Diversity of Ecosystem Governance'. *Global Studies Quarterly* 3(2). https://doi.org/10.1093/isagsq/ksad028

Maglia, Cristiana and Elana Wilson Rowe. 2024. 'Ecosystems and Ordering: A Dataset'. *Data in Brief*: 111085.

Martin, Lisa L. and Beth A. Simmons. 1998. 'Theories and Empirical Studies of International Institutions'. *International Organization* 52(4): 729–757.

McDonald, Matt. 2013. 'Discourses of Climate Security'. *Political Geography* 33: 42–51.

McDonald, Matt. 2021. *Ecological Security*. New York: Cambridge University Press.

Milkoreit, Manjana et al. 2024. 'Governance for Earth System Tipping Points: A Research Agenda'. *Earth System Governance* 21: 100216.

Musgrave, Paul. 2021. 'Political Science Has Its Own Lab Leaks'. *Foreign Policy*. https://foreignpolicy.com/2021/07/03/political-science-dangerous-lab-leaks/

Newell, Peter and Matthew Paterson. 2010. *Climate Capitalism: Global Warming and the Transformation of the Global Economy*. Cambridge: Cambridge University Press.

Oels, Angela. 2012. 'From "Securitization" of Climate Change to "Climatization" of the Security Field: Comparing Three Theoretical Perspectives', in Jürgen Scheffran, Michael Brzoska, Hans Günter Brauch, Peter Michael Link and Janpeter Schilling (eds) *Climate Change, Human Security and Violent Conflict*. Heidelberg: Springer, pp 185–205.

Paes, Lucas de Oliveira. 2022. 'The Amazon Rainforest and the Global–Regional Politics of Ecosystem Governance'. *International Affairs* 98(6): 2077–2097.

Paes, Lucas de Oliveira. 2023. 'Networked Territoriality: A Processual–Relational View on the Making (and Makings) of Regions in World Politics'. *Review of International Studies* 49(1): 53–82.

Paterson, Matthew and Michael Grubb. 1992. 'The International Politics of Climate Change'. *International Affairs* 68(2): 293–310.

Pattberg, Philipp and Oscar Widerberg. 2015. 'Theorising Global Environmental Governance: Key Findings and Future Questions'. *Millennium: Journal of International Studies* 43(2): 684–705.

Pereira, Joana C. and Eduardo Viola. 2020. 'Close to a Tipping Point? The Amazon and the Challenge of Sustainable Development under Growing Climate Pressures'. *Journal of Latin American Studies* 52(3): 467–494.

Pikialasorsuaq Commission. 2017. 'People of the Ice Bridge: The Future of the Pikialasorsuaq'. http://pikialasorsuaq.org/en/Resources/Reports

Raustiala, Kal and David G. Victor. 2004. 'The Regime Complex for Plant Genetic Resources'. *International Organization* 58(2): 277–309.

Reddekop, Jarrad. 2022. 'Against Ontological Capture: Drawing Lessons from Amazonian Kichwa Relationality'. *Review of International Studies* 48(5): 857–874.

Rockström, Johan et al. 2023. 'Safe and Just Earth System Boundaries'. *Nature* 619(7968): 102–111.

Rockström, Johan et al. 2009. 'Planetary Boundaries: Exploring the Safe Operating Space for Humanity'. *Ecology and Society* 14(2): 1–32.

Rockström, Johan et al. 2023. 'Safe and Just Earth System Boundaries'. *Nature* 619: 102–111.

Simangan, Dahlia. 2020. 'Where Is the Anthropocene? IR in a New Geological Epoch'. *International Affairs* 96(1): 211–224.

Sterling-Folker, Jennifer. 2015. 'All Hail to the Chief: Liberal IR Theory in the New World Order'. *International Studies Perspectives* 16(1): 40–49.

Sylvester, Christine. 2013. 'Experiencing the End and Afterlives of International Relations/Theory'. *European Journal of International Relations* 19(3): 609–626.

Wilson Rowe, Elana. 2021. 'Ecosystemic Politics: Analyzing the Consequences of Speaking for Adjacent Nature on the Global Stage'. *Political Geography* 91: 102497.

Yao, Joanne. 2019. '"Conquest from Barbarism": The Danube Commission, International Order and the Control of Nature as a Standard of Civilization'. *European Journal of International Relations* 25(2): 335–359.

Yao, Joanne. 2021. 'An International Hierarchy of Science: Conquest, Cooperation, and the 1959 Antarctic Treaty System'. *European Journal of International Relations* 27(4): 995–1019.

Yao, Joanne. 2022. *The Ideal River: How Control of Nature Shaped the International Order*. Manchester: Manchester University Press.

Young, Oran R. 1989. *International Cooperation: Building Regimes for Natural Resources and the Environment*. Ithaca, NY: Cornell University Press.

2

Governing Peace and Security in the Anthropocene

Dahlia Simangan

Introduction

The Anthropocene is a geological age that is proposed to replace the current Holocene epoch. Although a group of researchers at the International Commission on Stratigraphy in March 2024 rejected the proposal to officially adopt the term (Witze 2024), the concept remains significant in academic disciplines and policy discourses. For scholars, it is a widely known and loosely used term to denote an era in which humans are the dominant force behind geological change. It has also entered the language of multilateral organizations, such as the United Nations (UN). In 2020, the UN Development Programme (UNDP) released its annual Human Development Report, *The Next Frontier: Human Development and the Anthropocene*, presenting policy perspectives that situate human development within the complexities of the Anthropocene. It represents the growing recognition that threats to security and development are no longer confined to state action and militarization. Environmental degradation and ecological imbalance create sources of insecurity – food insecurity, forced displacement, infectious diseases, economic shocks and resource scarcity, to name but a few. The Report calls for ecologically aligned norms, values and institutions to balance socioeconomic and ecological wellbeing. It is a timely reminder of humanity's collective agency in creating and addressing global problems, thereby infusing a sense of urgency and responsibility in dealing with them.

The popularity of the Anthropocene as a universal term for the 'age of humans' comes with inherent problems. For one thing, it can potentially ignore the 'progressive impacts of humans on the world' (Subramanian 2019). A fixed geological marker fails to account for the historical and ongoing impacts of human activities on global environmental systems (Grear 2015). This could be why a scientific consensus on when it began is yet to be reached. Whether the Anthropocene started as early as our mastery of fire or as recently as the Great Acceleration, the different forms of global environmental change we are experiencing now can be traced to at least one of those markers, signifying the cumulative impact of human activities on the Earth system. Relatedly, the generalizing tendencies of the term do not adequately reflect and could potentially reinforce the inequalities and hierarchies within human societies (Grear 2015). As such, several scholars proposed alternative terms that better capture the heterogeneity of human history in relation to our planet (Haraway 2015; Moore 2016; Grove 2018; Pyne 2020). Despite, or perhaps because of, the controversies of such a term, it is imperative to scrutinize the conceptual boundaries of the Anthropocene. As Baskin (2015) suggests, the tensions and questions the term brings with it must be grappled with if academic disciplines wish to draw insights from its radical potential. The international relations (IR) discipline has started engaging with the Anthropocene discourse, particularly about the challenges and possibilities of global environmental governance. However, questions as to how existing governance mechanisms can solve global environmental issues leave out conversations on how addressing these issues would bring changes to global politics and governance (Beaumont et al, Introduction, in this volume). In other words, addressing the environmental challenges in the Anthropocene first necessitates addressing the issues of governance – shifting our thinking from the governance of nature to transforming governance for nature. This chapter contributes to this shift by examining the implications of the Anthropocene for governing peace and security.

In order to govern peace and security in the Anthropocene, it is imperative to rethink vulnerability, militarism and growth as these entrench the anthropocentric, state-centric and linear characteristics of the world order. The Anthropocene reveals nature-culture entanglement, the porousness of territoriality and the complexity of our systems. Aided by critical IR discourses on the Anthropocene, this chapter aims to inform more ecologically aligned implementation of reconciliation, security and development in postconflict or conflict-affected societies. Table 2.1 summarizes the three premises that structure the discussion in this chapter. First, the anthropocentric nature of the world order necessitates a rethinking of vulnerability in the context of postconflict reconciliation, in consonance with the entanglement of nature and culture. Second, militarism, as a form of security, primarily serves state interests and appears incompatible with the porousness of territoriality, as the Anthropocene reveals. Third, a linear approach to progress needs to be revisited as it presents growth as inherent and desirable in achieving development, reinforcing linearity that offers no alternative future(s) (Fagan 2019). Following these premises, this chapter

Table 2.1: Three premises for rethinking and governing peace and security in the Anthropocene

Premise	Prevailing worldviews in the world order	It is imperative to rethink these concepts vis-à-vis peace and security	The Anthropocene reveals
1	Anthropocentrism	Vulnerability (reconciliation)	Nature–culture entanglement
2	State-centrism	Militarism (security)	Porousness of territoriality
3	Linearity	Growth (development)	Complexity of systems

presents recent developments in governing peace and security in response to the challenges in the Anthropocene.

The prevailing worldviews in world order

The IR discipline has been instrumental in understanding world order. Realist and liberal framings dominated the field until the 1990s, with constructivist interjections as global interconnections intensified. Meanwhile, the study of global governance highlights the limitations of the theories of its parent discipline in explaining power and agency beyond the purview of states. More recently and as a response to potential disruptions in the world order, including the COVID-19 pandemic and the ongoing climate crisis, several IR scholars have started analysing the complexities, contingencies and relationalities that influence global politics, economics, as well as the environment (Rothe et al 2021). The practical and ontological questions surrounding climate change and other manifestations of global environmental transformation led to scholarly debates challenging some of the foundations of global governance (see McDonald in Chapter 3 in this volume). These debates engage with the concept of the Anthropocene, despite its universalist assumptions, as a useful backdrop for magnifying the consequences of preserving anthropocentric, state-centric and linear worldviews for governing the world. Indeed, the scientific and philosophical questions about the Anthropocene prompted a paradigm shift across disciplines and even policy making (for example, Hamilton 2015; Maslin and Lewis 2015; UNDP 2020; Biermann 2021), including politics and IR (Baskin 2015; Mitchell 2017; Dryzek and Pickering 2019).

Modern history, from settler colonialism to the emergence of nation states and the globalization of capital, was instrumental to shaping the worldviews that underpin prominent IR theories, such as (and in particular) realism and liberalism (Wallerstein 2011; Buzan and Lawson 2015; Bell 2019). However, the relevance of these worldviews is being questioned in the Anthropocene debates. Among these are the nature-culture binary, the rigidity of political borders and the linearity of human history. Building on my earlier thinking about the incompatibilities of these worldviews with the type of governance needed in the Anthropocene (Simangan 2020, 2024a), I argue that the anthropocentric, state-centric and growth-driven characteristics of the world order are some of the underlying pathological conditions that led to this new epoch.

The first is anthropocentrism. International politics is largely an extension of our political selves. To borrow Abraham Lincoln's prose about democracy from his Gettysburg address, the world order is 'of the people, for the people, and by the people'. And this anthropocentric political ordering has led to many modern conveniences that, to some extent, served the way in which we govern peace and security. This ordering rests on the treatment of nonhuman nature as ineligible subjects of politics, serving only human interests. This logic, Mitchell (2014) argues, is enshrined in the concept of human security, in which the human is the ultimate object of harm and the most important subject of security. This framing becomes problematic upon the realization that humans inflict harm not only to each other but also to other nonhuman beings. Echoing Mitchell (2014: 5) again, 'harm does not happen to humans in isolation, but rather to worlds composed of diverse beings'. Moreover, the triple planetary crisis that threatens our very own survival implies that anthropocentric projections on international politics delay the diagnosis of the causes of global environmental issues. The politics of IR should not be limited to the remit of human activity, especially when 'the structuring power of anthropocentric assumptions creates outcomes that adversely affect some of nature,', as Youatt (2014: 209) put it. For example, on the topic of solar radiation management (SRM) to reduce the impacts of climate change, there is a growing discussion on its economic, political and security implications (Dalby 2015; Biermann and Möller 2019; Symons 2019). International politics has not yet fully integrated the natural world (Corry and Kornbech 2021), and geoengineering and other ecomodernist solutions could unintentionally reassert human centrality and domination of nature. Further, a blind faith in technological innovations could potentially enable the preservation of a fossil fuel-based global economy (Gunderson et al 2020). The posthumanist turn in IR has already demonstrated the value of drawing on human-nature entanglement and relationality when dealing with the ethical questions about governing the climate (Cudworth and Hobden 2011; Eroukhmanoff and Harker 2017). While it is understandable that we cannot escape our 'humanness', it is not so when we think of political life as linked with nonhuman nature. Recognizing the limits of anthropocentric politics in confronting the complex web of harms and being harmed is not a disavowal of human triumphs, but an openness to more humble alternatives.

The second prevailing characteristic of the world order, which is also rooted in anthropocentrism, is state-centrism. While the study of global governance is broad enough to include nonstate actors, the system upon which it rests is still largely at the behest of states. The role of states remains central in political organizations, such as the UN, as well as most of the economic, legal and environmental regimes that influence the behaviour of other global actors. While there have always been criticisms that national interest is amorphous and capricious and that state-centric theories are limited to fully account for the influence of transnational flows and actors at both the global and domestic levels, states remain central actors and therefore important units of analysis in IR (Lake 2007; Buzan and Lawson 2015). In international policy negotiations, state-centric priorities, values and practices can prevent or enable agreements. In the context of peace and security, the political playbook still features the principle of non-intervention in the interests of state sovereignty. Despite significant steps in aligning peace and security imperatives with human rights and human security, there are still incidents in which state interests supersede the need to protect civilian population – from the military junta's refusal to allow the entry of humanitarian workers in the aftermath of Cyclone Nargis in Myanmar (Özerdem 2010), to Russia's claim that humanitarian assistance to Syria is a breach to the latter's sovereignty (Averre and Davies 2015) and, more recently, to the indiscriminate attacks against noncivilian populations caught in the war between Hamas and Israel, which are tantamount to 'genocidal warfare' (Semerdjian 2024: 1), on the grounds of rights to self-determination and self-defence (Heinze 2024; Samuel 2023). The same state-centrism, juxtaposed with growth-driven development, which will be explained later on in this chapter, has prevailed in international discussions over environmental problems. The UN, for one, traditionally approached environmental issues through law among nations and socioeconomic development, and failed to consider the peace and rights components of global environmental governance, further marginalizing vulnerable populations (Conca 2015). This blindspot enabled 'powerful actors to pursue destructive private agendas that undercut the common good' (Conca 2015: 12). These powerful actors, including some high-income countries and multinational corporations, exert influence in climate negotiations and their subsequent implementation. For example, while the negotiation stalemate surrounding climate financing finally ended when the establishment of the loss and damage fund was officially announced at the 2023 Conference of Parties, uncertainties remain about the sufficiency, transparency and accountability of the pledges (Editorial 2023; Gibson 2023; South 2024). Although pressures from civil society will likely be a factor, hopefully advancing the progress of delivering the pledges, ultimately, it is up to the state parties and their financial regimes to ensure that these pledges will reach the vulnerable groups in countries most affected by climate change.

The third characteristic is linearity, associated with hierarchical and static approaches to ordering the world. Linearity assumes a direct causal relationship between events, in isolation from other factors. Such thinking has influenced the theorization in IR; for example, democratization will lead to peace, peace interventions to political stability and economic growth to development, among other simplistic cause-and-effect assumptions. This reductionist view ignores multiple interacting factors and fails to account for historical contexts, path dependencies and unpredictable outcomes. Linearity is incompatible with the nature of global politics, which, as with any other holistic system, is subject to constant change. Power has diffused through the emergence of nonstate actors and nontraditional security issues that challenge the state-system and macro-level institutions (Rosenau 1990). Despite the growing recognition of the shortcomings of linear thinking in IR scholarship (for example, Rosenau 1990; Jervis 1997; Kavalski 2015), linearity remains at the helm of foreign policies and security strategies – from economic sanctions to territorial competition and nuclear deterrence. Discussions about climate change also adopt a similar fashion – that is, more research and better science will lead to more certainty, harmonization of interests and rational policies (Beck 2011). These linear assumptions are detached from the social, political and historical contexts of quantifiable anthropogenic greenhouse gas emissions and have favoured mitigation over adaptation measures (Beck 2011). In the context of the climate-conflict nexus, it is also important to account for nonlinearities. For example, in Africa, climate adaptation strategies in one locality could cause violence in another, and improved resilience to one disaster might not necessarily lead to the same outcome in a different type of disaster (Cappelli et al 2023). These spatial spillovers reflect the dynamic and self-organizing nature of complex systems and emphasize the need to recognize feedback loops and uncertainties.

The Anthropocene helps reveal how these characteristics reinforce each other. Anthropocentrism, state-centrism and linearity are intertwined in their reliance on and reinforcement of each other. For example, anthropocentric views on vulnerability prioritize security apparatuses that primarily serve anthropocentric and state-centric interests, failing to incorporate the role of and their impact on nonhuman nature. Human hubris refuses unpredictable outcomes and the capacity of complex systems to self-organize. Relatedly, the linearity of growth-driven development policies buttresses the instrumentalization of nonhuman nature for human purposes. Like linearity, territoriality further bounds politics and governance, despite spatial and temporal variations in natural and social processes. The causes and challenges in the Anthropocene also demonstrate the inadequacies of these views and approaches in governing global environmental change, as well as the need for a kind of governance that aligns with nature–culture entanglement, the porousness of territoriality and complexity.

The Anthropocene as a paradigm shift in international relations

Although the Earth is as animate as human beings, most of humanity, for much of its history, has been deanimating the Earth (Latour 2014). The entanglement of nature and culture, which remains in several Indigenous ontologies and animist traditions, was gradually eroded by a humanist version of modernity and humanist processes of globalization (Simangan 2019; see also Chakrabarty 2009). In the Anthropocene, anthropogenic impact on the rest of the physical world and the inadequacies of anthropocentric measures to incorporate feedbacks and spillovers erase this nature-culture dichotomy. As in one of Chakrabarty's theses, the anthropogenic causes of climate change collapse the distinction between natural history and human history (Chakrabarty 2009). Recognizing the problems of human-nature dualism must also be coupled with welcoming other ways of knowing about our place in nature. For example, the Yanomami practices of listening to voices from stories, poems, myths, dreams and songs allow them to coexist with the nonhuman world (Inoue 2018). While IR has traditionally favoured positivist approaches to knowing, interpretivist approaches could make sense of complexity – what is made and remade (Neufeld 1993). Furthermore, the challenges in the Anthropocene call for humanist scholarship to converse with 'deep time' scientists to develop nonanthropocentric perspectives (Chakrabarty 2020) and for the co-production of knowledge with other 'worlds' (Inoue 2018) – worlds of diverse forms of agencies and relationalities – other than the universalizing world of only people and markets (Escobar 2016).

Another manifestation of modernity in IR is territoriality. If the world order is continuously being remade, fixed territoriality will fall behind political transformations. To understand these transformations, Ruggie (1993) developed the notion of 'unbundled territoriality', taking into account nonterritorial forms of social organization and tracing back to how social organizations defined territories and not the other way round. Despite debates on either the persistence or obsolescence of territorial states whenever the world order is confronted by unprecedented global challenges, some IR theories remained fixated on the primacy of states and therefore territoriality and sovereignty (Agnew 1994). Bounded territoriality also informs contemporary geopolitics and international norms based on the premise that 'states represent nations' and culture and identity 'can be at least mapped into bounded territories' (Dalby 2020: 56). Postmodernism has helped interrogate the 'boundaries and boundedness' of territoriality, not to prescribe a conceptual framework, but to encourage theoretical plurality and not to call for the end of borders, but to consider 'new modes of inclusion and exclusion' (Albert 1998: 63–65; see also Ruggie 1993). Climate change has revived several ways of Indigenous thinking about territoriality and identity. Small island states of the Pacific and Indian Oceans do not identify as 'small islands' because of their territorial landmass, but rather as 'large ocean states' teeming with marine biodiversity (Chan 2018). The porousness of territoriality is the unbundling of sovereignty from the state and of culture from a place. The Anthropocene prompts a rethinking beyond state-bound and place-based territoriality towards relationality within and of worlds.

Relationality is also linked to complexity. As mentioned earlier in this chapter, complexity theory had already found its way into IR, challenging ontological and epistemological conceptions of agency and linearity while acknowledging 'uncertainties, dilemmas, and paradoxes' in social systems (Kavalski 2007: 40). 'Complexity theory describes the characteristics and functions of a particular type of holistic system that has the ability to adapt and that demonstrates emergent properties, including self-organizing behavior' (de Coning 2020: 2). More recently, it has also informed studies on the climate-conflict nexus (Beaumont and de Coning 2022). In contrast to the relatively stable Holocene, the Anthropocene is an epoch of Earthly ruptures – of shrinking spatialities and conflating temporalities. The sixth mass extinction is looming, inciting anxieties about uncertainties relating to survival, meaning and morality (Simangan 2023). This is an existential question for a discipline, such as IR, that has historically favoured humanist problem solving and predicated stability. How can the IR discipline approach the challenge of securing humanity from itself while dealing with the declining relevance of its anthropocentric, state-centric and linear frameworks? IR alone is not equipped to grasp the complexity of social systems in a changing Earth system. In a 'world of multiple disciplines' (Corry 2022: 291), it must harness multiple perspectives through transdisciplinary research and dialogue. Complex adaptive thinking aligns with perspectives calling for theoretical pluralism (for example, Albert 1998; Ferguson 2015; Beaumont and de Coning 2022). Such thinking is useful for shifting away from determinist traditions towards decolonized, emancipatory and global IR.

Rethinking peace and security in the Anthropocene

These paradigm shifts prompted by the Anthropocene have implications for peace and security. The relationship between peace or conflict on the one hand and environment and climate on the other is well established in academic and policy circles. Early studies demonstrate the link between conflict and the environment through issues of competition over scarce or abundant resources, the environmental impact of war, weapons and military activities, even during peacetime, and the similarities of war and environmental degradation in their destructive and transboundary nature (Brock 1991; see also Le Billon 2001). Conflicts are also understood to intensify due to the impacts of climate change by exacerbating existing conditions of competition and violence in conflict-affected societies where institutions are usually ineffective, public services are insufficient and people are agriculturally dependent (Koubi 2019; see also Mach et al 2019). This relationship is by no means direct and linear, as political, economic and social factors mediate their influence over each other. Following this recognition, peacebuilding efforts across scales have started integrating environmental considerations and vice versa. The UN, for example, in several reports and proclamations, recommends conflict-sensitive climate change adaptation policies and supports climate-sensitive peacebuilding efforts (for example, UN 2015; UN Department of Political and Peacebuilding Affairs, nd). Despite these developments, there are concerns that increasingly unprecedented anthropogenic impact on the biosphere will likely provoke unknown geopolitical risks, especially in regions with increasing competition over transboundary resources and where military tensions and nuclear threats are high (Subramanian 2019). Climate change-induced displacement and migration, for instance, challenge existing governance institutions (Burkett 2011; Warner 2010) and even our very notion of what constitutes a state (Fainu 2023), especially given the rapid development of digital technologies. Without underestimating the role of environmental cooperation in easing these tensions (Ide et al 2021), the changing Earth system will have adverse implications for peace and security.

The paradigm shifts following the debates surrounding the Anthropocene could inform how we rethink and govern peace and security in this age of complex and entangled uncertainties. I aim to contextualize these governance imaginaries in postconflict peacebuilding, which is understood as the 'construction of a new environment ... which seeks to avoid the breakdown of peaceful conditions' (UN 1992: 15). It is a multidimensional process of building and sustaining peace in societies transitioning from conflict. More specifically, I argue that the pursuits of reconciliation, security and development – the conventional components of peacebuilding – could benefit from more ecologically aligned worldviews.

First, rethinking vulnerability is a shift from anthropocentric reconciliation towards interspecies justice. When the environment becomes a vehicle for war and conflict – for example, in scorched earth tactics and control over natural resources – animals, trees and nonhuman nature also fall victim to violence. They can be incidental, for example, due to toxic pollution (Atherton 2021), or direct victims, for example, being shot live (Weinstein

2010). However, the environment is rarely considered in postconflict reconciliation, failing to recognize and repair the harm done to nonhuman nature. Vulnerability is generally conceived as a human experience. In justice and reconciliation mechanisms – from South Africa to Rwanda and Timor-Leste – reparations are meant for the human victims of war and violence. While these mechanisms were valuable in documenting abuses, healing trauma, reconciling groups and preventing similar tragedies, the non-inclusion of nonhuman subjects is limiting. Wars would have looked different without the objectification of nonhuman animals for tactical strategies and experimentation and, more positively, for uplifting the morale of combatants (Cudworth and Hobden 2015). Furthermore, despite community-led reconciliation efforts, many of these mechanisms were facilitated or dependent on international organizations or state institutions (Glucksam 2024). These characteristics of reconciliation mechanisms evoke anthropocentrism and state-centrism.

Echoing the posthumanist strand of critical IR scholarship, rethinking vulnerability beyond the human species not only challenges state-centrism and anthropocentrism, but also allows the possibility of recognizing interspecies violence and promoting interspecies reconciliation. Some works on this issue emphasize human-nature entanglement in global politics and acknowledge the personhood and rights of nonhuman nature (for example, Youatt 2017; Burke and Fishel 2019; Pereira and Saramago 2020). In framing cosmopolitan belongingness or interspecies togetherness, Leep (2018: 65) presents it as a 'pathway for looking back and forward, pursuing a responsibility informed by the past with a sense of anticipatory belongingness'. This framing also reflects the conflation of temporalities as magnified in critical Anthropocene discourse in IR, and the inclusion of nonhuman subjects broadens the scope of agencies and vulnerabilities, prompting a rethinking of security. For example, converging animal ethics and humanitarian law would require modifications in military culture and international norms (Milburn and van Goozen 2023). Incorporating animals into the ethics of war broadens the scope of reconstruction, rehabilitation and memorialization to include environmental efforts, such as reforestation and wildlife conservation (Milburn and van Goozen 2023). Such convergence also relates to questions regarding the relevance of militarism amid global environmental change.

The second and related proposal for governing peace and security in the Anthropocene is to rethink militarism, shifting state-centric security and bounded territoriality towards relationality. Mann (1987) defines militarism as 'a set of attitudes and social practices which regards war and the preparation for war as a normal and desirable social activity'. From the US declaration of the 'war on terror' in 2001 to the ongoing Ukraine-Russia and Israel-Hamas wars, militarism is a prevailing characteristic of global politics. War is commonly understood as an extension of politics (Clausewitz 2003) and maintaining internal peace is tied to the potential for war against external actors (Kapferer 2004). The perpetual threat of military invasion in early European history led to the conception of territorial integrity and state sovereignty, becoming the principles for modern IR. When the state apparatus is undermined or challenged, waging war against external or internal forces becomes a legitimate option. In addition to high levels of military expenditure, militarization of domestic social relations and tendencies to use force,[1] the nuclear arms race is central to militarism (Smith 1983). Optimists predicted that the end of the Cold War would make nuclear weapons obsolete and that innovations in cyberwarfare would serve as countermeasures to nuclear acquisition. However, as in the words of UN Secretary-General António Guterres (2022), 'the end of the Cold War also left us with a dangerous falsehood: that the threat of nuclear war was a thing of the past'. Nuclear proliferation is rooted in the state's anticipation of threats and compounded by its strategic interests, security commitments and nature of alliances (Monteiro and Debs 2014; Reiter 2014). The increasing complexity and plurality of security issues brought about by rapid technological advancements and the advent of a multipolar world order, among other power dynamics, led many to believe that the 'Third Nuclear Age' is upon us. This age is characterized by great power nuclear competition, regional conflicts and the emergence of new nuclear powers – a potent combination for inciting conflict and nuclear crises in a world of 'First Nuclear Age politics with Second Nuclear Age technology' (Futter and Zala 2021: 274).

While it seems that militarism, kept alive by state-centric interests, is here to stay, it is becoming out of place when pursuing peace and security in postconflict contexts. There is a breadth of literature on the importance of relationality in peacebuilding (for example, Hunt 2017; Söderström et al 2021; Torrent 2021). Peacebuilding that lacks local ownership and is insensitive to local contexts can revive underlying causes of violence or create new ones. These critiques have led to more inclusive and locally attuned approaches to peacebuilding, including the concepts of the local turn (Mac Ginty and Richmond 2013), hybrid peace (Boege et al 2009; Mac Ginty 2010) and everyday peace (Firchow 2018; Mac Ginty 2021). This relational turn in peacebuilding is also aligned with environmental peacebuilding. Several environmental initiatives among local stakeholders in conflict-affected societies, such as in Yemen, Ghana and Timor-Leste, have encouraged peaceful coexistence (Taher et al 2012; Bukari et al 2018; Ide et al 2021). Ignoring these community-led initiatives could alienate the people from the peacebuilding process, resulting in further environmental degradation and conflict risks. Therefore, rather than prioritizing state-centric security apparatuses, such as military capabilities, during postconflict reconstruction, governing peace and security in the Anthropocene invests in supporting relational interactions between peoples and with their environment. While this sounds utopian, a security discourse brought about by relational processes of negotiation and contestation, and one that includes all living beings, both at present and in the future – or what McDonald (2021) calls ecological security – is not only ethically desirable but also politically possible and, in addition, urgently actionable.

Another ethically desirable and politically possible way for governing peace and security is rethinking growth as a precursor of development. Like militarism, growth also has expansionist tendencies. Capital is the engine of growth, and globalization made capital the currency of economic affairs. Through a combination of coercion and persuasion, states' economic policies are crafted to achieve high levels of gross domestic product (GDP), which has become the benchmark for prosperity and growth (Fioramonti 2017). A growth-driven economy rests on production, accumulation and consumption, often resulting in environmental degradation and socioeconomic inequalities. These inequalities are elevated and even magnified at the international level as states without a dominant position in securing energy and resources remain disadvantaged from the division and movement of 'labor, capital, resources, and waste' (Lengefeld and Smith 2013: 14). While economic growth under capitalism indeed contributed to improved human conditions, it also objectified skillsets and commodified social interactions (Sewell Jr. 2014). At the same time, in Foster's (2022: 26) words, 'the logic of capital accumulation, which accepts no boundary beyond itself, now threatens in the era of catastrophe capitalism the very nature of existence on earth'. Overproduction and overconsumption exacerbate inequalities across levels. A growth-driven economy incentivizes those with resources to produce and consume more than necessary while depriving others of their basic needs (Simangan 2024b). At the global level, and using waste trade as an example, Global South countries have become the final destination of waste produced by Global North countries. Trade liberalization and existing disparities in GDP levels and environmental standards facilitate waste trade – argued as a form of ecological imperialism – easing the disposal burden of exporting countries while causing social and environmental repercussions for importing countries (Cotta 2020). Waste trade and other manifestations of inequalities within the broader global political economy are byproducts of a growth-driven economy. Economic growth is not a foolproof path towards development, especially the type of development that is cognizant of the need to overcome the environmental challenges in the Anthropocene.

There are compelling reasons to shift from a linear, growth-driven development to complex adaptive approaches. It is a long-held assumption in postconflict development that economic growth is essential to peace and security. Indeed, in several cases, economic growth driven by foreign trade, aid and investment improved public services and addressed sources of insecurity. However, the unrealistic pursuit of sustained economic growth could incentivize unsustainable consumption, resource exploitation and unequal distribution of resources, especially in conflict-affected or postconflict contexts where political and social institutions to manage these detrimental impacts are ineffective or nonfunctional. Several scholarly works examined the alternatives to the growth principle to advance wellbeing and maintain an ecologically sustainable society (for example, Nicoson 2021; Hasselbalch et al 2023; Fioramonti 2024; see also Bosi-Moreira and Kranke in Chapter 7 in this volume). Decentralized, localized and care economies are examples of just, participatory and ecologically sustainable ways of meeting wellbeing needs while promoting social cohesion. Instead of promoting economic growth in postconflict societies, these postgrowth alternatives respond to inequalities and injustices heightened in the Anthropocene (Simangan 2024b).

From governance of nature to the transformation of governance for nature

The UN recognizes natural resource depletion as a source of global insecurity and recommends climate change adaptation policies that are sensitive to conflict contexts (UN 2015). This implies that governing environmental issues prompts reforms of existing practices and institutions. The Anthropocene brings multilateralism concerning peace and security to a critical juncture. The first UN Agenda for Peace was published in 1992, when the international community was transitioning from a bipolar world order maintained by nuclear power rivalries. In response to nonmilitary and nonstate threats to peace and security, it became the foundation for multilateral action on international peace and security. The strength of this multilateralism is being challenged by the persistence and evolving nature of armed conflicts, climate crisis and increasing inequalities within and among nations, among other global issues. These issues cannot be resolved with the growing distrust and division among nations. As the UN Secretary-General reminds us: 'We are now at an inflection point. The post-Cold War period is over. A transition is under way to a new global order. While its contours remain to be defined, leaders around the world have referred to multipolarity as one of its defining traits' (UN 2023: 3). And it is imperative for actors of multilateralism to reorient their actions.

The 2023 New Agenda for Peace is one of the opportunities for new directions in governing peace and security (UN 2023). It lists several priority areas for international cooperation, which has recently seen an erosion since the Russian invasion of Ukraine and, more recently, with the war between Israel and Hamas. One of these priority areas is climate emergency: 'Where record temperatures, erratic precipitation and rising sea levels reduce harvests, destroy critical infrastructure and displace communities, they exacerbate the risks of instability, in particular in situations already affected by conflict' (UN 2023: 6). To address the links between climate, peace and security, the UN Secretary-General urges, among other things, the creation of an expert group within the Intergovernmental Panel on Climate Change (IPCC) dedicated to developing climate-sensitive peacebuilding approaches. He also recommends establishing regional hubs on climate, peace and security to better contextualize and connect technical solutions and policies, as well as a climate-dedicated funding within the UN Peacebuilding Fund. Relatedly, the 28th Conference of Parties (COP) for the United Nations Framework Convention on Climate Change (UNFCCC) in 2023 was the first COP to feature Recovery and Peace as a thematic convening day to enhance support for climate action in countries or communities experiencing conflict or fragility. This commitment echoes the recognition that conditions of conflict and fragility exacerbate climate vulnerabilities and reiterates the importance of ensuring that climate adaptation programmes do not incite tensions or contribute to conflict. This is welcome news, considering the gap in climate adaptation funding for conflict-affected and fragile states (Crisis Group 2022). It signals a new momentum towards policy convergence in climate action and peace.

The New Agenda also echoes previous attempts to transform approaches to peace and security in accordance with environmental considerations. In 2007, scientists behind the Doomsday Clock, a symbolic representation of humanmade global destruction, decided to include climate change as a source of this destruction. The Clock used to focus only on threats from nuclear weapons, but our understanding of what constitutes risks to humanity has expanded because of climate change (Bronson 2023). In the same vein, peace and security risks are no longer within the remit of the social or the political; they are also linked to the environment, particularly the human impact on the biosphere. This expansion suggests that security is temporary – even more so socially constructed and politicized – from the perspective of human agency, temporality and spatiality. Governing peace in the Anthropocene deems demilitarization as either unnecessary or detrimental to human and ecological security. Therefore, the New Agenda's call for nuclear disarmament aligns with a peace agenda in the Anthropocene, which demands and supports strategies towards demilitarization.

Policy convergence around climate and conflict risks is critical to governing peace in the Anthropocene. While it is understandable that the Secretary-General's transformative vision will unlikely find favour among some conservative and powerful member states (Gowan 2023b), the New Agenda can be considered a flagship document for preparing the international community for a new set of challenges to peace and security in the context of global environmental change. Despite the absence of specific proposals aimed at creating new multilateral institutions for addressing global problems (Gowan 2023a), some conceptualizations and recommendations in the New Agenda can prompt further discussions about a type of multilateralism fit for a changing global environment. If current multilateralism aims to confront the challenges in the Anthropocene, it must build on these signals of transformation.

Conclusion

How should peace and security be governed in an age of global transformations? In this chapter, I retraced some prevailing characteristics of the world order that are incompatible with the entangled agencies, shrinking spatialities and conflating temporalities in the Anthropocene. These characteristics, as operationalized in multilateral responses to peace and security, reinforce practices and institutions with a long history of marginalizing nonhuman nature, resorting to military responses and exporting market economies. As a result, most responses to complex and cascading global environmental changes are bound by short-term security priorities and narrow geopolitical interests contradicting the profound transformations the Anthropocene demands. However, despite its humanist foundations, the IR discipline has contributed to critical interrogations of anthropocentric vulnerability, military-centric security and growth-driven development by drawing on theoretical pluralism and methodological diversity. The discipline's engagement with the Anthropocene discourse infuses a renewed sense of urgency to understand and rethink responses to the new challenges to peace and security.

The challenges in the Anthropocene are profound and necessitate transformative disruptions. Governing peace and security in the Anthropocene welcomes the creation and reordering of institutions and practices that recognize the vulnerabilities of all beings, support relational security imperatives and address complex wellbeing. It is holistic, encompassing all forms and structures of violence, and inclusive of nonhuman or more-than-human agencies. To govern peace and security amid challenges in the Anthropocene is to transform governance into one that is intergenerational, multispecies and planetary.

Note

[1] These manifestations of militarism remain true to this day with the rise in global military expenditures (Tian et al 2024), successful military coups in several countries around the world, more prominently in Africa (Ero and Mutiga 2024), and the growing acceptance of the use of force in peacekeeping operations and humanitarian interventions.

References

Agnew, J. (1994) 'The territorial trap: the geographical assumptions of international relations theory', *Review of International Political Economy*, 1(1): 53–80.

Albert, M. (1998) 'On boundaries, territory and postmodernity: an international relations perspective', *Geopolitics*, 3(1): 53–68.

Atherton, K.D. (2021) 'U.S. forces are leaving a toxic environmental legacy in Afghanistan', *Scientific American*. Available at: https://www.scientificamerican.com/article/u-s-forces-are-leaving-a-toxic-environmental-legacy-in-afghanistan/

Averre, D. and Davies, L. (2015) 'Russia, humanitarian intervention and the responsibility to protect: the case of Syria', *International Affairs*, 91(4): 813–834.

Baskin, J. (2015) 'Paradigm dressed as epoch: the ideology of the Anthropocene', *Environmental Values*, 24(1): 9–29.

Beaumont, P. and de Coning, C. (2022) 'Coping with complexity: toward epistemological pluralism in climate–conflict scholarship', *International Studies Review*, 24(4): viac055.

Beck, S. (2011) 'Moving beyond the linear model of expertise? IPCC and the test of adaptation', *Regional Environmental Change*, 11(2): 297–306.

Bell, D. (2019) *Reordering the World: Essays on Liberalism and Empire*. Princeton: Princeton University Press.

Biermann, F. (2021) 'The future of "environmental" policy in the Anthropocene: time for a paradigm shift', *Environmental Politics*, 30(1–2): 61–80.

Biermann, F. and Möller, I. (2019) 'Rich man's solution? Climate engineering discourses and the marginalization of the Global South', *International Environmental Agreements: Politics, Law and Economics*, 19(2): 151–167.

Boege, V., Brown, M.A. and Clements, K.P. (2009) 'Hybrid political orders, not fragile states', *Peace Review*, 21(1): 13–21.

Brock, L. (1991) 'Peace through parks: the environment on the peace research agenda', *Journal of Peace Research*, 28(4): 407–423.

Bronson, R. (2023) 'Measuring existential risk', *Peace Policy: Solutions to Violent Conflict*, 55: 6–7.

Bukari, K.N., Sow, P. and Scheffran, J. (2018) 'Cooperation and co-existence between farmers and herders in the midst of violent farmer-herder conflicts in Ghana', *African Studies Review*, 61(2): 78–102.

Burke, A. and Fishel, S. (2019) 'Power, world politics and thing-systems in the Anthropocene', in E. Lövbrand and F. Biermann (eds) *Anthropocene Encounters: New Directions in Green Political Thinking*. Cambridge: Cambridge University Press, pp 87–108.

Burkett, M. (2011) 'The nation ex-situ: on climate change, deterritorialized nationhood and the post-climate era', *Climate Law*. https://doi.org/10.1163/CL-2011-040

Buzan, B. and Lawson, G. (2015) *The Global Transformation: History, Modernity and the Making of International Relations*. Cambridge: Cambridge University Press.

Cappelli, F. et al (2023) 'Climate change and armed conflicts in Africa: temporal persistence, non-linear climate impact and geographical spillovers', *Economia Politica*, 40(2): 517–560.

Chakrabarty, D. (2009) 'The climate of history: four theses', *Critical Inquiry*, 35(2): 197–222.

Chakrabarty, D. (2020) 'The human sciences and climate change: a crisis of anthropocentrism', *Science and Culture*, 86(1–2): 46–48.

Chan, N. (2018) '"Large Ocean States": sovereignty, small islands and marine protected areas in global oceans governance', *Global Governance*, 24(4): 537–555.

Clausewitz, C. (2003) *On War*. London: Penguin.

Conca, K. (2015) *An Unfinished Foundation: The United Nations and Global Environmental Governance*. Oxford: Oxford University Press.

Corry, O. (2022) 'What's the point of being a discipline? Four disciplinary strategies and the future of international relations', *Cooperation and Conflict*, 57(3): 290–310.

Corry, O. and Kornbech, N. (2021) 'Geoengineering: a new arena of international politics', in D. Chandler, F. Müller and D. Rothe (eds) *International Relations in the Anthropocene: New Agendas, New Agencies and New Approaches*. Cham: Springer International Publishing, pp 95–112.

Cotta, B. (2020) 'What goes around, comes around? Access and allocation problems in Global North–South waste trade', *International Environmental Agreements: Politics, Law and Economics*, 20(2): 255–269.

Crisis Group (2022) 'Giving countries in conflict their fair share of climate finance'. Available at: https://www.crisisgroup.org/content/fair-share-of-climate-finance

Cudworth, E. and Hobden, S. (2011) *Posthuman International Relations: Complexity, Ecologism and Global Politics*. London: Zed Books.

Cudworth, E. and Hobden, S. (2015) 'The posthuman way of war', *Security Dialogue*, 46(6): 513–529.

Dalby, S. (2015) 'Geoengineering: The next era of geopolitics?', *Geography Compass*, 9(4): 190–201.

Dalby, S. (2020) *Anthropocene Geopolitics: Globalization, Security, Sustainability*. Ottawa: University of Ottawa Press.

De Coning, C. (2020) 'Insights from complexity theory for peace and conflict studies', in O.P. Richmond and G. Visoka (eds) *The Palgrave Encyclopedia of Peace and Conflict Studies*. Cham: Springer, pp 1–10.

Dryzek, J.S. and Pickering, J. (2019) *The Politics of the Anthropocene*. Oxford: Oxford University Press.

Editorial (2023) '"Loss and damage": the most controversial words in climate finance today', *Nature*, 623: 665–666.

Ero, C. and Mutiga, M. (2024) 'The crisis of African democracy: coups are a symptom – not the cause – of political dysfunction essays', *Foreign Affairs*, 103(1): 120–134.

Eroukhmanoff, C. and Harker, M. (eds) (2017) *Reflections on the Posthuman in International Relations: The Anthropocene, Security and Ecology*. E-International Relations. Available at: https://www.e-ir.info/wp-content/uploads/2017/09/Reflections-on-the-Posthuman-in-IR-E-IR.pdf

Escobar, A. (2016) 'Thinking-feeling with the Earth: territorial struggles and the ontological dimension of the epistemologies of the South', *Revista de Antropología Iberoamericana*, 11(1): 11–32.

Fagan, M. (2019) 'On the dangers of an Anthropocene epoch: geological time, political time and post-human politics', *Political Geography*, 70: 55–63.

Fainu, K. (2023) 'Facing extinction, Tuvalu considers the digital clone of a country', *The Guardian*, 27 June. Available at: https://www.theguardian.com/world/2023/jun/27/tuvalu-climate-crisis-rising-sea-levels-pacific-island-nation-country-digital-clone

Ferguson, Y.H. (2015) 'Diversity in IR theory: pluralism as an opportunity for understanding global politics', *International Studies Perspectives*, 16(1): 3–12.

Fioramonti, L. (2017) *The World after GDP: Politics, Business and Society in the Post Growth Era*. Cambridge: Polity Press.

Fioramonti, L. (2024) 'Post-growth theories in a global world: a comparative analysis', *Review of International Studies*, 50(5): 866–876.

Firchow, P. (2018) *Reclaiming Everyday Peace: Local Voices in Measurement and Evaluation after War*. Cambridge: Cambridge University Press.

Foster, J.B. (2022) *Capitalism in the Anthropocene: Ecological Ruin or Ecological Revolution*. New York: Monthly Review Press.

Futter, A. and Zala, B. (2021) 'Strategic non-nuclear weapons and the onset of a third nuclear age', *European Journal of International Security*, 6(3): 257–277.

Gibson, S. (2023) Don't applaud the COP28 climate summit's loss and damage fund deal just yet – here's what's missing, *The Conversation*. Available at: http://theconversation.com/dont-applaud-the-cop28-climate-summits-loss-and-damage-fund-deal-just-yet-heres-whats-missing-218093

Glucksam, N. (2024) 'The many conceptions of post-conflict reconciliation: learning from practitioners', *Peacebuilding*. https://doi.org/10.1080/21647259.2024.2370686

Gowan, R. (2023a) 'How the UN can make the most of the new agenda for peace', *International Crisis Group*. Available at: https://www.crisisgroup.org/global/how-un-can-make-most-new-agenda-peace

Gowan, R. (2023b) 'What's new about the UN's New Agenda for Peace?', *International Crisis Group*. Available at: https://www.crisisgroup.org/global/whats-new-about-uns-new-agenda-peace

Grear, A. (2015) 'Deconstructing anthropos: a critical legal reflection on "anthropocentric" law and anthropocene "humanity"', *Law and Critique*, 26(3): 225–249.

Grove, J.V. (2018) 'The geopolitics of extinction: from the Anthropocene to the Eurocene', in D.R. McCarthy (ed.) *Technology and World Politics: An Introduction*. Abingdon: Routledge, pp 204–223.

Gunderson, R., Stuart, D. and Petersen, B. (2020) 'The fossil fuel industry's framing of carbon capture and storage: faith in innovation, value instrumentalization, and status quo maintenance', *Journal of Cleaner Production*, 252: 119767.

Guterres, A. (2022) 'Nuclear weapons are not yesterday's problem, they remain today's growing threat'. Available at: https://www.un.org/sg/en/content/sg/articles/2022-01-04/nuclear-weapons-are-not-yesterday%E2%80%99s-problem-they-remain-today%E2%80%99s-growing-threat

Hamilton, C. (2015) 'Getting the Anthropocene so wrong', *The Anthropocene Review*, 2(2): 102–107.

Haraway, D. (2015) 'Anthropocene, capitalocene, plantationocene, chthulucene: making kin', *Environmental Humanities*, 6(1): 159–165.

Hasselbalch, J.A., Kranke, M. and Chertkovskaya, E. (2023) 'Organizing for transformation: post-growth in International Political Economy', *Review of International Political Economy*, 30(5): 1621–1638.

Heinze, E.A. (2024) 'International law, self-defense, and the Israel-Hamas conflict', *Parameters*, 54(1): 71–86.

Hunt, C.T. (2017) 'Beyond the binaries: towards a relational approach to peacebuilding', *Global Change, Peace & Security*, 29(3): 209–227.

Ide, T. et al (2021) 'The past and future(s) of environmental peacebuilding', *International Affairs*, 97(1): 1–16.

Ide, T., Palmer, L.R. and Barnett, J. (2021) 'Environmental peacebuilding from below: customary approaches in Timor-Leste', *International Affairs*, 97(1): 103–117.

Inoue, C.Y.A. (2018) 'Worlding the study of global environmental politics in the Anthropocene: Indigenous voices from the Amazon', *Global Environmental Politics*, 18(4): 25–42.

Jervis, R. (1997) *System Effects: Complexity in Political and Social Life*. Princeton: Princeton University Press.

Kapferer, B. (2004) 'Introduction: Old permutations, new formations? War, state and global transgression', *Social Analysis: The International Journal of Social and Cultural Practice*, 48(1): 64–72.

Kavalski, E. (2007) 'The fifth debate and the emergence of complex international relations theory: notes on the application of complexity theory to the study of international life', *Cambridge Review of International Affairs*, 20(3): 435–454.

Kavalski, E. (2015) *World Politics at the Edge of Chaos: Reflections on Complexity and Global Life*. New York: SUNY Press.

Koubi, V. (2019) 'Climate change and conflict', *Annual Review of Political Science*, 22(1): 343–360.

Lake, D.A. (2007) 'The state and international relations'. https://doi.org/10.2139/ssrn.1004423

Latour, B. (2014) 'Agency at the time of the Anthropocene', *New Literary History*, 45(1): 1–18.

Le Billon, P. (2001) 'The political ecology of war: natural resources and armed conflicts', *Political Geography*, 20(5): 561–584.

Leep, M. (2018) 'Stray dogs, post-humanism and cosmopolitan belongingness: interspecies hospitality in times of war', *Millennium Journal of International Studies*, 47(1): 45–66.

Lengefeld, M.R. and Smith, C.L. (2013) 'Nuclear shadows: weighing the environmental effects of militarism, capitalism, and modernization in a global context, 2001–2007', *Human Ecology Review*, 20(1): 11–25.

Mac Ginty, R. (2010) 'Hybrid peace: the interaction between top-down and bottom-up peace', *Security Dialogue*, 41(4): 391–412.

Mac Ginty, R. (2021) *Everyday Peace: How So-Called Ordinary People Can Disrupt Violent Conflict*. Oxford: Oxford University Press.

Mac Ginty, R. and Richmond, O.P. (2013) 'The local turn in peace building: a critical agenda for peace', *Third World Quarterly*, 34(5): 763–783.

Mach, K.J. et al (2019) 'Climate as a risk factor for armed conflict', *Nature*, 571(7764): 193–197.

Mann, M. (1987) 'The roots and contradictions of modern militarism', *New Left Review*, 1(162).

Maslin, M.A. and Lewis, S.L. (2015) 'Anthropocene: Earth system, geological, philosophical and political paradigm shifts', *The Anthropocene Review*, 2(2): 108–116.

McDonald, M. (2021) *Ecological Security: Climate Change and the Construction of Security*. Cambridge: Cambridge University Press.

Milburn, J. and van Goozen, S. (2023) 'Animals and the ethics of war: a call for an inclusive just-war theory', *International Relations*, 37(3): 423–448.

Mitchell, A. (2014) 'Only human? A worldly approach to security', *Security Dialogue*, 45(1): 5–21.

Mitchell, A. (2017) 'Is IR going extinct?', *European Journal of International Relations*, 23(1): 3–25.

Monteiro, N.P. and Debs, A. (2014) 'The strategic logic of nuclear proliferation', *International Security*, 39(2): 7–51.

Moore, J.W. (2016) *Anthropocene or Capitalocene? Nature, History, and the Crisis of Capitalism*. Oakland, CA: PM Press.

Neufeld, M. (1993) 'Interpretation and the "science" of international relations', *Review of International Studies*, 19(1): 39–61.

Nicoson, C. (2021) 'Towards climate resilient peace: an intersectional and degrowth approach', *Sustainability Science*, 16(4): 1147–1158.

Özerdem, A. (2010) 'The "responsibility to protect" in natural disasters: another excuse for interventionism? Nargis cyclone, Myanmar', *Conflict, Security & Development*, 10(5): 693–713.

Pereira, J.C. and Saramago, A. (2020) *Non-human Nature in World Politics: Theory and Practice*. New York: Springer Nature.

Pyne, S.J. (2020) 'From pleistocene to pyrocene: fire replaces ice', *Earth's Future*, 8(11): e2020EF001722.

Reiter, D. (2014) 'Security commitments and nuclear proliferation', *Foreign Policy Analysis*, 10(1): 61–80.

Rosenau, J.N. (1990) *Turbulence in World Politics: A Theory of Change and Continuity*. Princeton: Princeton University Press.

Rothe, D., Müller, F. and Chandler, D. (2021) 'Introduction: international relations in the Anthropocene', in D. Chandler (ed.) *International Relations in the Anthropocene: New Agendas, New Agencies and New Approaches*. Cham: Palgrave Macmillan, pp 1–16.

Ruggie, J.G. (1993) 'Territoriality and beyond: problematizing modernity in international relations', *International Organization*, 47(1): 139–174.

Samuel, M.T. (2023) 'The Israel-Hamas war: historical context and international law', *Middle East Policy*, 30(4): 3–9.

Semerdjian, E. (2024) 'A world without civilians', *Journal of Genocide Research*: 1–6. https://doi.org/10.1080/14623528.2024.2306714

Sewell Jr., W.H. (2014) 'The capitalist epoch', *Social Science History*, 38(1–2): 1–11.

Simangan, D. (2019) 'Situating the Asia Pacific in the age of the Anthropocene', *Australian Journal of International Affairs*, 73(6): 564–584.

Simangan, D. (2020) 'Can the liberal international order survive the Anthropocene? Three propositions for converging peace and survival', *The Anthropocene Review*, 9(1): 37–51.

Simangan, D. (2023) 'How should IR deal with the "end of the world"? Existential anxieties and possibilities in the Anthropocene', *Review of International Studies*, 49(5): 855–871.

Simangan, D. (2024a) 'Peace in the Anthropocene', *Oxford Research Encyclopedia of International Studies*, 50(5): 877–887.

Simangan, D. (2024b) 'Post-growth peacebuilding', *Review of International Studies*: 1–11.

Smith, R. (1983) 'Aspects of militarism', *Capital & Class*, 7(1): 17–32.

Söderström, J., Åkebo, M. and Jarstad, A.K. (2021) 'Friends, fellows, and foes: a new framework for studying relational peace', *International Studies Review*, 23(3): 484–508.

South, D.W. (2024) 'Loss and damage fund – operationalized at COP28 but funding and allocation process unresolved', *Climate and Energy*, 40(7): 29–32.

Subramanian, M. (2019) 'Anthropocene now: influential panel votes to recognize Earth's new epoch', *Nature*. https://doi.org/10.1038/d41586-019-01641-5

Symons, J. (2019) 'Realist climate ethics: promoting climate ambition within the classical realist tradition', *Review of International Studies*, 45(1): 141–160.

Taher, T. et al. (2012) 'Local groundwater governance in Yemen: building on traditions and enabling communities to craft new rules', *Hydrogeology Journal*, 20(6): 1177–1188.

Tian, N. et al. (2024) 'Trends in world military expenditure, 2023'. *SIPRI*. Available at: https://www.sipri.org/publications/2024/sipri-fact-sheets/trends-world-military-expenditure-2023

Torrent, I. (2021) *Entangled Peace: UN Peacebuilding and the Limits of a Relational World*. Lanham, MD: Rowman & Littlefield.

UN (United Nations) (1992) *An Agenda for Peace: Preventive Diplomacy, Peacemaking and Peace-Keeping*. UN Doc. A/47/277.

UN (2015) 'The challenge of sustaining peace: report of the advisory group of experts for the 2015 review of the United Nations peacebuilding architecture'. Available at: https://reliefweb.int/report/world/challenge-sustaining-peace-report-advisory-group-experts-2015-review-united-nations

UN (2023) *Our Common Agenda Policy Brief 9: A New Agenda for Peace*. Available at: https://dppa.un.org/en/a-new-agenda-for-peace

UN Department of Political and Peacebuilding Affairs (nd) 'Addressing the impact of climate change on peace and security, political and peacebuilding affairs'. Available at: https://dppa.un.org/en/addressing-impact-of-climate-change-peace-and-security

UNDP (United Nations Development Programme) (2020) *Human Development Report 2020. The Next Frontier: Human Development in the Anthropocene*. New York: UNDP.

Wallerstein, I. (2011) *The Modern World-System I: Capitalist Agriculture and the Origins of the European World-Economy in the Sixteenth Century*. Berkeley: University of California Press.

Warner, K. (2010) 'Global environmental change and migration: governance challenges', *Global Environmental Change*, 20(3): 402–413.

Weinstein, A. (2010) 'Iraq's slumdog massacre: one million dogs face death'. Available at: https://www.motherjones.com/politics/2010/06/iraq-kbr-one-million-dogs-death/

Witze, A. (2024) 'Geologists reject the Anthropocene as Earth's new epoch – after 15 years of debate', *Nature*, 627(8003): 249–250.

Youatt, R. (2014) 'Interspecies relations, international relations: rethinking anthropocentric politics', *Millennium: Journal of International Studies*, 43(1): 207–223.

Youatt, R. (2017) 'Personhood and the rights of nature: the new subjects of contemporary Earth politics', *International Political Sociology*, 11(1): 39–54.

3

The Anthropocene, Climate Change and (Ecological) Security

Matt McDonald

Introduction

Security is a master concept of IR thought and practice. The promise of providing it underpins the legitimacy of key institutions of global politics, whether states or international organizations. For the former, the commitment to provide security is central to the social contract. For the latter, the commitment to the 'maintenance of international peace and security' is featured as Article 1 of the United Nations (UN) Charter. And for prominent theories of international relations, the quest for security drives key dynamics of global politics and attempts to govern or regulate it. But to what extent do either the practices and governance of security in global politics, or theories of security in IR thought sufficiently come to terms with the scale of contemporary ecological crises or the broader Anthropocene context in which these are embedded? Does the embrace of climate change as a security issue in states or international organizations point to a (necessary) shift in the ways in which climate change, security and/or global governance are conceived or approached?

This chapter explores these questions, pointing in the process to the problematic logics that continue to inhere in prominent approaches to 'climate security' even when this link is acknowledged or embraced. Of course, that there are limits to existing global practices and governance arrangements in effectively coming to terms with the realities of climate change is a conclusion that may be drawn on the basis of the onward march of the realities of climate change, species extinction and unprecedented (in the human experience at least) challenges to ecosystems and their resilience. But simply pointing to the failure of states and the broader 'international community' to address a challenge like climate change in the 40 years in which we have been aware of the problem doesn't get us to the reasons for this failure or means of addressing it. Rather, coming to terms with this failure requires examining the core assumptions and logics underpinning IR thought and global practice regarding the Anthropocene and ecological crises. Here, the assumptions, logics and pathologies of 'security' provide a particularly important insight into this failure, not least as a central concept in IR thought.

Ultimately, as this chapter will explore, even when we have seen recognition of the threats posed by climate change, this has stopped well short of enabling or ushering in the types of measures or practices necessary for genuinely addressing the harms associated with it. The engagement of the UN Security Council (UNSC) with climate change, for example, has increased over time and states involved in these deliberations are eager to affirm the importance of action to address the challenges posed by climate change. Yet a resolution recognizing the UNSC's responsibilities to contribute to addressing this threat – even involving minimal commitments such as regular reporting to the Secretary-General or the integration of climate considerations in the design and execution of peace operations, for example – has proven to be beyond the capacity or willingness of the institution or its member states. In this sense, and as contributors to this debate argued at the time (see McDonald 2023b), the capacity of the institution to claim to be fulfilling its role as the primary institution for 'maintaining international peace and security' – central to the UN mandate – could be called into question. Speaking to the core concerns of this volume, it is certainly possible to conclude that the inherent limitations of attempts to govern climate change responses through the lenses and institutions of security- for reasons that will be examined- risk further undermining global order and the legitimacy of key institutions within it.

In simple terms, the central idea or image of security evident in both theory and modes of security governance- that of self-contained units attempting to insulate themselves from deliberate external threats- is simply inimical to coming to terms with the challenge of an issue like climate change. In the context of climate change the origins of the 'threat' are predominantly unintended (and everyday) action both within and beyond the state; the state lacks the capacity to unilaterally address that threat regardless of its material power, capacity or contribution to the problem; the traditional instruments of (national) security are either largely marginal to the means of responding to it or even a significant part of the problem (see, for example, Crawford 2022); the idea of insulation or conservation of self-contained units does not get to grips with the reality of ongoing (and unavoidable) change to ecosystem functionality and with it life prospects; the necessity for cooperation between states immediately runs into complex ethical challenges and problematic governance systems, structures and norms; and, perhaps most fundamentally, the separation of humanity and nature central to conceptions of what it means to be secure – admittedly dominant in modern political thought (see Soper 1995; Clark and Szerszynski 2021) – is fundamentally at odds with the need to come to terms with the reality of the Anthropocene context and the idea of humanity as irrevocably embedded in the ecological contexts upon which it relies for its ongoing existence. All these points suggest fundamental impediments to addressing ecological crises through existing institutions and practices. They also suggest fundamental impediments to understanding the origins of these crises, or guiding responses to them, through the dominant frameworks of traditional IR thought. These limitations become evident when examining current institutional responses to the climate change–security relationship, and the international relations discourses with which these responses align.

This chapter is divided into three sections. In the first section, I outline engagement with security in IR thought and practice. I identify key assumptions regarding the meaning and logic of security in IR thought before examining what the practice or governance of security looks like. The second section outlines the challenge of the Anthropocene context and ecological crises to this image of security, focusing in particular on the challenge of climate change. After noting the scale of this challenge, I examine attempts in IR thought and practice to come to terms with the security implications of climate change. The third and final section reflects on where this analysis leaves us, pointing to the promise of approaching the security implications of ecological crises through the lens of ecological security, in which our focus turns to the resilience of ecosystems themselves. I situate this approach in broader attempts to challenge the human-nature divide, before noting key challenges facing this approach and its institutionalization that speak to the core concerns of this volume as a whole.

Security in international relations

Security is a master concept in IR thought and practice. Of course, for traditional approaches to international relations such as realism, the (endless) search for security – defined principally in terms of the survival of the state – is the driving force of IR. This has always been a moral project, as Michael Williams (1998) has argued. At the heart of the realist defence of the primacy of the state is the social contract: the idea that individuals sacrifice some degree of freedom and autonomy to the state – the Leviathan, in Hobbes' terms – in return for their protection in an anarchic and dangerous external environment. A concern with states and their security is, then, a concern with the welfare of people. Viewed in this light, it becomes easier to understand why state leaders consistently emphasize that the promise of providing security is the most important responsibility of the government itself. Perhaps more importantly for the purposes of this chapter and the broader text of which it is a part, recognizing this point gives us a sense of the centrality of security to IR thought and practice.

In the national security framework,[1] which is still dominant in IR thought and practice in the face of sustained critique from scholars in the 'critical security studies' tradition (see Browning and McDonald 2013), security is of – and for – states. It is the state whose security – defined in terms of sovereignty and territorial integrity – matters. And it is the state, through self-help and military preparedness in particular, that provides security. Building on its realist foundations, the imperative of self-help in a world characterized by the absence of a higher authority than states encourages the aggressive and single-minded pursuit of material power relative to others to enable the state's survival, and further facilitate the state's capacity to ensure that survival through the projection of power externally.[2]

The alignment of referent and agent of security in this framework provides a neat answer to key questions of ethical focus and responsibility, with the state positioned as being concerned with protecting itself. However, this necessarily positions the space of security practice as one in which self-contained units (states) attempt to insulate themselves from external threats to their sovereignty and territorial integrity. By extension, issues become security issues to the extent that they have the capacity to threaten or undermine sovereignty or territorial integrity. This establishes relatively clear criteria for considering whether issues of environmental change are considered on the national security agenda, but also encourages a highly partial view of the nature of the challenge posed by environmental change. While climate change causes significant and widespread harm to peoples and ecosystems, fundamentally undermines conditions for human survival and even challenges the capacity of states to function, it does not follow that – within the national security framework – it warrants consideration as a national security threat (Levy 1995). I will return to this point later in this chapter.

In the international security framework, in which the focus turns to the stability of the international system and/or the protection of the rules and norms of an international society (see Bellamy and McDonald 2004), there is clearly more room for states – still the most powerful actors – to work together to address threats to security. This capacity to recognize 'absolute' rather than simply relative gains allows for the possibility of addressing transnational threats, which is clearly more consistent with recognizing and responding to environmental change (see Garcia 2010). However, the question of referent object and agent of security becomes significantly more complicated. As critics have rightly noted, abstract notions of an 'international community' in need of either defending or acting appear at once both unclear and partial in their account of whose community matters or exactly who is responsible for acting (Buzan and Gonzalez-Pelaez 2005).

A more productive account of what international security means in IR theory might here draw on the English School focus on international society (Dunne 1998; Bellamy and McDonald 2004). Within this international society, we see a shifting emphasis (in terms of what constitute dominant norms and rules) between concerns with 'order' (pluralism) and 'justice' (solidarism). While the former orients towards the maintenance of the status quo and absence of disruption to that order, the latter focuses our attention on potentially responding to harms regardless of their effect on international stability (enabling practices that challenge sovereignty as non-intervention, such as humanitarian intervention). There is an implication that states can and should work together to address either the disruptive implications of the effects of environmental change that threaten international stability or to address the harmful effects of environmental change through both prevention and response. As such, states remain the most powerful agents, but recognize the need to work together in concert with other actors to address challenges like climate change.

In the national security framework, which is embedded in realist thought, the pathway for an issue like climate change to be considered a national security threat is limited (Levy 1995). If the requirement is the potential threat posed to sovereignty and territorial integrity, few states experience such considerations directly as a result of climate change. At best we might acknowledge the potential immediate threat posed to low-lying Pacific Island states, for example, whose territory (even sovereignty by some accounts) is immediately and directly threatened by rising sea levels associated with the effects of climate change (Maas and Carius 2012; Mitra and Sanghi 2023). But this scenario is rare, and for the most part our focus in the national security framing is with the secondary implications of climate change for national security: the threat posed by population movements or conflict itself linked to or caused by climate change (Busby 2007, 2008). Indeed, this is at the heart of the 'threat multiplier' (CNA 2007, 2014) framing that has become dominant in accounts of the national security implications of climate change and its potential relationship to armed conflict. Rather than a threat in and of itself, climate change (despite the existential challenge it demonstrably represents) becomes a threat only when linked to the potential for it to trigger armed conflict. The state responds to these indirect challenges in turn not primarily through prevention (mitigation), but by addressing the consequences of this challenge through adaptation (McDonald 2013). As will be noted in the following section, while we see variations in different states' engagement with the (national) security implications of climate change, in practice this framework is evident in the way in which states conceive of the key dimensions of threat and response.

In the case of an international security framework, the nature of the threat posed by climate change is still predominantly an indirect one: the potential for the direct effects of climate change (warmer temperatures, changing rainfall patterns, desertification, ocean acidification or disasters) to precipitate *genuine* threats associated with large-scale population movement, disruptive state fragility, regional instability or armed conflict. This is arguably the lowest common denominator vision of an international security focus – the absence of large-scale and disruptive instability. Indeed, it is significant to note here that the 'threat multiplier' framework has been invoked by a range of international organizations and in the context of debates within the UN Security Council about the international security implications of climate change (Hardt 2021; Maertens 2021; Scartozzi 2022; McDonald 2023b). There is a potential here – evident in the UN Security Council debates noted previously – for states and international organizations to push in the direction of a focus on harm, potentially emphasizing the role of mitigation action as a means of advancing security. Yet, as will be noted later on, such conceptualizations remain marginal to institutional engagement with the security implications of climate change, whether at the national or the international level.

Thus, in traditional accounts of security in IR, the story is ultimately one of key institutions of global politics attempting to insulate themselves from the destabilizing effects of external developments. In the dominant national security account, states are concerned with protecting their own sovereignty and territorial integrity from these external threats. In the case of international security states might be capable of working together to advance or address (transnational) threats that pose a challenge to international order and stability, but the threat remains largely indirect, practices orient towards the responsive (that is, adaptation), and responsibility for action depends in this schema on the extent to which we see institutional recognition of a threat (McDonald 2024a). Neither of these frameworks appears to be well suited to making sense of or guiding responses to the nature of ecological crises we face in the Anthropocene epoch. And the disinclination of key institutions to move beyond these dominant accounts risks undermining their own legitimacy in the face of the onset of climate change and, with it, potentially undermining the broader global order.

Security, the ecological challenge and the Anthropocene

It is difficult to do justice to the scale of contemporary ecological crises, memorably characterized in Chapter 1 of this volume as 'omni-crises'. Recent prominent attempts to identify and track challenges to 'planetary boundaries', for example, demonstrate a consistent failure to address biosphere integrity, chemical pollution, ocean acidification, freshwater availability, land system change, nitrogen and phosphorous flows to oceans and the biosphere, atmospheric aerosol loading and, of course, climate change (Rockström et al 2009, 2023). The impact of humanity on the natural world is such that 'human-made mass now exceeds all living biomass' (Elhacham et al 2020). According to a range of accounts, we are experiencing a 'sixth mass extinction' event – a 'biological annihilation' (Ceballos, Ehrlich and Dirzo 2017) – with a recent report suggesting a 70 per cent decline in wildlife populations in the last 50 years (WWF 2022). And then there's climate change.

While often presented as a pressing future challenge, the reality of climate change and its effects have shifted to the here and now. The most recent IPCC (2022) report noted that average global temperatures were already 1.1°C over pre-industrial levels, with consistent spikes over the Paris target of 1.5°C evident over the course of 2023. The climate change already experienced has resulted in more frequent and more severe disasters, species loss, changing rainfall patterns, desertification, rising sea levels and ocean acidification. This has already had devastating effects on ecosystems, some of which (for example, coral reefs) are vulnerable to relatively low-level temperature rises. With the goal of preventing dangerous warming looking less attainable, international discussions about the response to the climate crisis have shifted. From an early exclusive focus on mitigation, topics and practices such as adaptation, resilience building, reparations and even geoengineering have joined mitigation as topics in formal international climate deliberations (Bulkeley and Newell 2023; McDonald 2023a).

In making sense of these varied harms and the impact of humanity as a force on the planet, a range of analysts have made a case for the arrival of the Anthropocene epoch. Paul Crutzen (2006: 16) argued that given 'major and still growing impacts of human activities on earth and atmosphere, and at all, including global, scales, it thus is more than appropriate to emphasise the central role of mankind in geology and ecology by using the term "Anthropocene" for the current geological epoch'.

The argument for the Anthropocene followed the logic that because human activities had so fundamentally affected the functionality of Earth systems, the central contribution of humanity to the contemporary geological epoch should be acknowledged (see also Steffen 2021). In this schema, humanity, simply put, is a planetary force profoundly affecting the conditions of our own existence.

The concept of the Anthropocene has been subject to critique for eliding crucial differences regarding *which* humans have driven ecological destruction or misrepresenting the particular set of forces that have been behind this process. Indeed some grounds of critique are noted in the Chapter 1 of this volume. Jairus Grove (2019) makes a case for the Eurocene in locating the origins of crisis in a European model of extraction and development, while Jason Moore (2016) makes a case for the 'capitalocene' in attributing responsibility to the pursuit of particular modes of economic exchange.[3] However, for these potential limitations, this concept does draw our attention to the impossibility of separating humanity from the ecological conditions of its existence. Ultimately, it is a framing that represents a fundamental challenge to the human-nature distinction that is viewed as central to modern thought (Soper 1995). Clark and Szerzynski (2021) describe this challenge as a fundamental one to the social sciences more broadly. So just how well equipped is IR as a discipline, and the institutions of global politics themselves, for coming to terms with this challenge and guiding effective responses to it? At least in the context of more traditional approaches to security focused on the preservation of national or international security, and assessed against the responses of key institutions to this challenge, the answer is not particularly well.

The particular challenges posed by climate change, and especially the necessity of particular forms of response, clearly sit uneasily with the dominant logics of security in IR thought and extant institutional arrangements and practices in global politics. Returning to the themes outlined at the start of this chapter, the specific *nature of the threat* climate change poses is the first fundamental challenge to note. The threat is clearly an existential one, with the capacity to undermine the long-term existence of life on the planet. Yet its origins in everyday and (predominantly) unintentional actions are inconsistent with the image of the origins of threat in intentional and deliberate action by an external adversary (see Walt 2021). This also speaks to the challenge of the Anthropocene context to modern political thought and, in particular, the idea of humanity as irrevocably embedded within the ecological contexts upon which it relies for its survival. Again, an 'insulation or protection from' model of security positions the threat in terms that are inconsistent with its essence, while also reaffirming a problematic conception of who it is that is to be secured from that threat. Indeed, for some, the logic of security – of protecting or conserving the status quo, whether at the domestic or the international level – clearly pushes against the reality of ongoing change that climate change represents (see Dalby 2015; Hardt et al 2024).

Following this point, the conception of agency and means of providing security in traditional accounts of security in IR thought, and dominant practices of international politics, is also a key limitation. States are clearly still imbued with primary responsibility for addressing climate change within the UNFCCC process, and are more broadly viewed as the key security actors in global politics, with a monopoly on the legitimate use of force. Yet, demonstrably, states lack the capacity to unilaterally address the threat posed by climate change regardless of their material power, capacity or contribution to the problem. And here the traditional instruments of (national) security – the military and defence sectors – are either of limited immediate utility in addressing the problem of climate change itself or actively contribute to it through their own emissions (see, for example, Crawford 2022; Depledge 2023; Vogler 2024).

Finally, the need for effective global cooperation to address a global challenge clearly runs into challenges in the form of dominant modes of governance and, as noted earlier, the challenge of facilitating and pursuing meaningful action cooperatively in a world of states. The range of acutely complex ethical and political challenges in formulating a fair agreement between almost 200 states clearly suggests that the contemporary state-based order would not be a starting point of choice for managing a genuinely global challenge like climate change. These include different vulnerabilities, capacity to address implications or adapt; reliance on or profit from the fossil fuel economy; cultural and social investment in vulnerable spaces or destructive activities; historical responsibility; and different political systems, among other things (see Gardiner 2013; Bulkeley and Newell 2023). And that's before considering obligations to other living beings or future generations. The failure of the UNFCCC process, established in 1992, to prevent dangerous climate change points directly to these limitations. And while we have seen increasing recognition of the security implications of climate change among states, this has generally not encouraged the type of reorientation of perspective or new practices consistent with addressing the scale of the threat faced.

Debates within the UN Security Council seem to bear out the limitations of IR frameworks and their application to contemporary global politics. Certainly, debates about climate change within the UNSC have been increasing in frequency since 2018 and with majority support for recognizing a role for the UNSC in addressing the security implications of climate change. But substantive suggestions/proposals have been limited (to information gathering and incorporation of climate considerations into peace operations, for example) and even then we saw the rejection of this resolution (with Russia using its veto power and both China and India expressing opposition) when formally put to the UNSC in December 2021 (see McDonald 2023b). For an institution that is widely viewed as bearing primary responsibility for the provision and maintenance of international peace and security – central to its mandate – the failure to accept responsibility to act as agents for the provision of security in the case of fundamental and existential threat suggests significant limitations to existing security governance structures in global politics.[4] Something isn't working.

These limitations are evident in practical attempts to recognize, grapple with and institutionalize responses to the security implications of climate change within states. A significant majority of states now recognize the security implications of climate change (see Vogler 2023, 2024), but even among those states that have moved to recognize and institutionalize a position on the security implications of climate change, the same problematic pathologies, logics and implications largely remain. In terms of the nature of the threat and the definition of the referent object of security, and with a handful of exceptions,[5] the dominant focus in policy statements is on the indirect or secondary implications of climate change for sovereignty, territorial integrity and the defence sector itself (see Hardt et al 2023). Following this, we see more overarching focus in such documents on the *external* sources of threat or challenge to nation states, including the potential for militaries to be drawn into new sites of conflict or humanitarian assistance and disaster relief (HADR) missions. And very few states actually recognize the importance of mitigation, with militaries – central agents of national security within states – usually exempted from government-wide emissions reduction targets (see Depledge 2023). In all of these senses, the institutions and practices of global politics most regularly adhere to the expectations of traditional theories of IR, militating against the type of change in perspective and approach necessary for coming to terms with the Anthropocene context and the scale of contemporary ecological crises. To return to the theme of this chapter – and the broader text – the inclination here is towards attempting to fit such crises within existing (security) governance arrangements.

Australia's recent (and belated) embrace of climate change as a security concern is illustrative here. In its most recent Defence Strategic Review (DoD 2023), for example, climate change was recognized as a threat to national security, which was hardly surprising given the exposure of Australia to disasters (such as bushfires) and the acute vulnerability of small island states in its Pacific region (see McDonald 2021). Yet here, discussion of climate change warranted barely one of 100 pages in this report; most of this content was making a case for why the military should be called upon *less* to respond to natural disasters associated with climate change, and while committing no additional funds or capacity to addressing climate change, the report proceeded to make a case for a multibillion dollar investment in nuclear-powered submarines to address traditional geostrategic rivalry concerns (Evans 2024). It is tempting to conclude that climate change was little more than a footnote to a business-as-usual national security statement, crowded out by other pressing threats or concerns. Arguably, however, the issue evident here is even more fundamental: a failure of the state to genuinely reckon with the nature of the challenge posed by climate change, or to reorient conceptions, priorities, institutions and practices in a manner that is consistent with addressing this existential threat. While the nature of the climate crisis should encourage a radical reorientation of global governance and practices, as this volume suggests, the approach of key institutions to the climate-security relationship suggests a determination to ensure climate change – at best – fits within comfortable accounts of self-contained entities insulating themselves from external threat (see Hardt et al 2024).

A way forward?

Given these evident limitations of dominant accounts and practices of security in IR, it is hardly surprising that many advise avoiding the security framing altogether. Some see a danger of militarization of environmental challenges through this framing (Deudney 1991), others simply view the logic of security as illiberal and pernicious (Wæver 1995), while for others, a security framing suggests conservation or protection rather than recognition of the need to come to terms with ongoing (environmental) change (Dalby 2015). It's hard to wholly reject these points; indeed, some have featured in objections raised by states to the discussion of climate change in the UNSC (see Maertens 2021). But as the contributors to this volume and an increasing volume of critical scholarship notes (see Burke et al 2016), it's harder to escape the conclusion that these limitations are evident in broader IR thought and practice as a whole. Here, the veneration of states and the state system, the liberal model of political economy and development, and the primacy of an image of security defined in terms of the absence of immediate threats to the sovereignty and territorial integrity of states all militate against effective action in response to ecological crises, and at worst serve to drive it. And if security is politically significant in defining core values, in underpinning the legitimacy of the key actors of global politics and/or in enabling particular (even exceptional) forms of response (see Browning and McDonald 2013), then it needs to be redefined and reformed rather than rejected (see Booth 2007).

There have been a range of attempts in IR thought to challenge traditional images of security associated with the preservation or conservation of (self-contained) institutions in the face of deliberate external threats (see Browning and McDonald 2013). While often questioning the ethical assumptions and implications of these choices, critical scholarship on security also draws our attention to the politics and significance of apparently abstract debates about whose security matters or what constitutes a threat to that referent object. Recognizing that the promise of providing security is central to the legitimacy of key institutions of global politics, even potentially enabling particular sets of responses to challenges designated as a threat (see Wæver 1995; Buzan et al 1998), illustrates what is at stake in debates about the meaning and scope of security. In the process, it points to the intersection between IR theory and the practices of global politics that both follow from and inform the conceptual frameworks we employ.

While a range of scholarship challenges traditional accounts of security – their assumptions and implications – on a variety of grounds, more specifically a number of accounts have engaged with the particular need to come to terms with the nature of ecological crises. Early articulations of the need to recognize the security implications of environmental change beyond the arbitrary (and unhelpful) borders of nation states invoked a 'one world' image of planetary or biosphere security (Myers 1993; Litfin 1999). More recently, some have developed and defended the concept of 'Anthropocene security' to recognize the embeddedness of human populations in ecological contexts (Harrington and Shearing 2017; Mobjörk and Lövbrand 2021). Others have challenged the human-nature divide through articulating defences of worldly security (Mitchell 2014), posthuman security (Cudworth and Hobden 2017) or ecospheric security (Welsh 2022), for example. But perhaps the most explicit articulation of an alternative framework for making sense of the security implications of climate change in particular is that of ecological security. While noting earlier invocations of ecological security (see, for example, Mische 1989; Barnett 2001; Pirages and Cousins 2005), my own elucidation of the concept (see McDonald 2021) defines it in terms of the resilience of ecosystems themselves in the face of the immediate and direct implications of climate change.

A number of elements of this approach to the climate-security relationship are important to draw out here as a response to the tendencies and limitations of traditional approaches to and practices of security in IR. First, on the referent object, the focus on ecosystems clearly shifts our emphasis away from the (moral) primacy of existing political institutions and towards the ecological. This orientation is viewed as the best means of encouraging sets of practices that prioritize the rights and needs of the most vulnerable. The emphasis on (ecosystem) resilience – rather than conservation or preservation – reflects the reality of ongoing (environmental) change, in turn reminding us of the origins of the resilience concept in ecological thought (see Bourbeau 2018). Second, on the nature of the threat, the focus here is on the immediate and direct harms caused by climate change – rather than secondary or indirect implications – to ecosystems and, by extension, the most vulnerable. Third, on means of response, the emphasis on direct and immediate harms associated with climate change in turn involves prioritizing urgent mitigation effort to address these harms. And finally, on agents or responsibility for addressing the security implications of climate change, those with capacity (to cause and/or address harms) are defined here as bearing responsibility to act on behalf of others (on these points, see McDonald 2021).

Such an approach clearly challenges traditional accounts of security – and security threats in the context of the Anthropocene and ecological crises – in profound ways. Rather than the image of self-contained units insulating themselves from (external and deliberate) threats, it suggests the need to orient towards recognizing the embeddedness of humanity in the ecological conditions of its own existence and the imperative of working to minimize (direct) harms caused by challenges like climate change. States – particularly wealthy states – bear responsibility for acting to advance ecological security through mitigation action, but are not the only actors obligated to act and are far from the focal point of moral concern in terms of climate impacts. It is a significant distance, in short, from dominant accounts and practices of security in IR.

However, this distance represents both a necessary correction and a significant (potential) weakness. While the case for the former has been made, the latter concerns its limited immediate purchase or resonance within corridors of power. The elegant simplicity of a national security framing – where the state is the agent of security providing its own security – provides immediate and clear buy-in (through self-interest) from states in terms of the imperative of acting to achieve security. This is clearly absent in the case of ecological security, wherein states are compelled to act for others, from vulnerable populations in other countries to future generations of humans and other living beings. Can we realistically expect such considerations to inform the way in which security is viewed by powerful actors in global politics? Should we be surprised that something approaching 'business-as-usual' security politics appears most common?

The record to date in the embrace of ecological security among states falls a long way short of providing us with cause for optimism. This is a core point in the comparative analysis of approaches to the climate-security relationship by member states of the UNSC, in which an ecological security focus is found to be distinctly marginal to such approaches (Hardt et al eds. 2023). Yet we do see movement in this direction in terms of state engagement, with some states compelling the military and defence sector towards mitigation (NZ MoD 2019) and many identifying the imperative of meaningful and concerted international action to (urgent) mitigation action. Indeed, in his analysis of the UK's approach to the relationship between climate change and security, Cameron Harrington (2023: 315) suggests that 'there are signs that an ecological security discourse is emerging that challenges the historic focus on climate change as an accelerator of traditional security threats, or as a threat to individual well-being in vulnerable states of the Global South'.

A key challenge in this context, for scholars of climate and security and in particular for advocates of ecological security, will be identifying the circumstances or contexts that might give rise to such an approach. Here, identifying and examining movement in the direction of ecological security – recognizing progress – is important in terms of allowing us to understand the potential for movement towards practices oriented towards the most vulnerable in the face of the climate crisis (see McDonald 2021: 188–191). Interviews I conducted with policy makers in states as distinct as the US, Germany, New Zealand and Sweden in 2022–2023 suggests core concerns, at least at particular times and for particular governments, with preventing immediate climate harms and addressing human security concerns in response to climate change (see McDonald 2024b). Significant drivers of this type of orientation that were identified by interview subjects included public expectations, the values and commitments of particular governments and the foreign policy identity of the state concerned (see McDonald 2024b).

Beyond this, and speaking to the core concerns of this volume, it may also be the case that addressing the immediate and direct harms (insecurity) of climate change necessitates and/or drives new institutional arrangements and modes of governance. Already scholars have pointed to the emergence of nontraditional forms of large-scale ecological governance (for example, Wilson Rowe 2021), the distinctive (and frequently effective) climate diplomacy of small states (for example, Morgan, Carter and Manoa 2024), to forms of localised governance in response to climate effects (for example, McNaught 2024) and to Indigenous forms of ecosystem management embedded in relational approaches to the natural world that challenge the human-nature divide (for example, Reed et al 2021). Ongoing grassroots activists also consistently make a case for prioritizing attention to the most vulnerable and addressing the direct harms of climate change (see, for example, Garcia-Gibson 2023). Indeed, Aly Tkachenko (2024) makes the case that existing forms of grassroots direct action to protect local environments can be viewed as ecological security in practice.

This reminds us of two important points in considering the prospects for ecological security and the dynamic relationship between nature governance and global order. First, we cannot allow existing institutional arrangements and their capacity – or preferred policy instruments – to define the limits of our conception of politics, sites or means of response (on this point, see Lövbrand et al 2015). Second, addressing climate change necessitates action below and above those traditional security providers – the state and its defence establishment. In this context, recognizing the nature of the challenge that climate change poses means recognizing the potential – even the imperative – of new forms of governance.

Conclusion

Traditional approaches to security in IR are ill-equipped to come to terms with the nature of the Anthropocene or the ecological crises – the 'omni-crises' – we face. Nor do such approaches orient towards *directly* addressing the threats posed by ecological crises, most notably climate change. This necessitates an orientation towards ecological security, which recognizes direct and immediate harms experienced by the most vulnerable, and prioritizes responses to these harms. Traditional accounts, by contrast, focus on the preservation or protection of self-contained units, and their insulation from the effects of an ecological crisis like climate change. This might be consistent with the logics or pathologies of (dominant accounts of) security, but it sits uneasily with the nature of this crisis. As noted in the preceding analysis, such an approach fits poorly with recognition of the immediate and direct harms associated with climate change, with the range of means and array of actors necessary to address it and, perhaps most fundamentally, struggle to come to terms with the core question of who is rendered most fundamentally and immediately vulnerable to the effects of climate change and ecological crises more broadly.

As the preceding analysis has also made clear, the tendencies within traditional theoretical approaches to IR are replicated and reinforced by the practices of key institutions of global politics. Here, even when institutions like states and the UNSC engage directly with the security implications of climate change, there is an overwhelming commitment to make climate change fit within existing accounts of the remit and scope of the organization and the logics of security with which they are comfortable. The result is – at best – the pursuit of piecemeal, partial and predominantly adaptive responses oriented towards the survival of the institution itself, not the fundamental shift in orientation and practice required to address the climate crisis. As the onset of the climate crisis deepens and worsens, the failure of business-as-usual conceptions and practices of security in response to that crisis will increasingly raise questions about the legitimacy of these institutions and also potentially the contours of global order.

This *tendency* in orientation does not, as noted in the preceding pages, apply to all scholarship or all institutions. At the institutional level, aside from increasing engagement with the security implications of climate change in the UNSC (see Maertens 2021; McDonald 2023b) and among states (see Vogler 2023, 2024), we can also see at least some movements away from the model of the self-contained state insulating itself from the (indirect) effects of climate change among those institutions, as noted by policy practitioners. In IR scholarship generally and security scholarship specifically, we can also see increasing recognition of the climate crisis and the Anthropocene context, along with reflection on what this means for changing the way we think about security and how it might be advanced or realized. This is evident in the critical scholarship on security generally and scholars making the case for ecological security – or variants of it – more specifically. In this sense, recognition of the realities of dealing with the climate crisis might yet drive changes in modes of global governance and practice, along with the conceptual frameworks used to make sense of these dynamics and guide responses to them. And there is arguably a role for scholars and scholarship attempting to advance this case in public debate, in the corridors of power and in the academy, a project to which this book contributes.

Notes

1. Although as Welsh (2022: 9) notes, it is more appropriate to position the referent object of security in this schema as the 'state' rather than the 'nation'.
2. Of course, for a critique of this 'logic' of anarchy – and the assumption that this compels self-help and mistrust – see Wendt (1992).
3. It should also be noted that the proposal to formally endorse the Anthropocene epoch, which dated from the 1950s, failed to receive a supermajority of votes from a panel of the International Commission on Stratigraphy in March 2024.
4. Indeed, Article 24 of the UN Charter notes that: 'In order to ensure prompt and effective action by the United Nations, its Members confer on the Security Council *primary responsibility* for the maintenance of international peace and security.'
5. The states of the Pacific Island Forum defined climate change primarily as a direct threat to human security in their 2018 Boe Declaration (PIF 2018), while policy statements of climate security in Sweden and Germany, for example, identify the immediate effects of climate change for vulnerable states in the developing world.

References

Barnett, Jon (2001) *The Meaning of Environmental Security: Ecological Politics and Policy in the New Security Era*. London: Zed Books.

Bellamy, Alex. and McDonald, Matt (2004) 'Securing international society: towards an English School discourse of security'. *Australian Journal of Political Science*, 39(2), 307–330.

Browning, Christopher and McDonald, Matt (2013) 'The future of critical security studies: ethics and the politics of security'. *European Journal of International Relations*, 19(2), 235–255.

Booth, Ken (2007) *Theory of World Security*. Cambridge: Cambridge University Press.

Bourbeau, Philippe (2018) *On Resilience: Genealogy, Logics and World Politics*. Cambridge: Cambridge University Press.

Burke, Anthony et al (2016) 'Planet politics: a manifesto from the end of IR'. *Millennium*, 44(3), 499–523.

Bulkeley, Harriet and Newell, Peter (2023) *Governing Climate Change*, 3rd edn. Abingdon: Routledge.

Busby, Joshua (2007) *Climate Change and National Security: An Agenda for Action*. New York: Council on Foreign Relations.

Busby, Joshua (2008) 'Who cares about the weather? Climate change and U.S. national security'. *Security Studies*, 17, 468–504.

Buzan, Barry, and Gonzalez-Pelaez, Ana (2005) '"International community" after Iraq'. *International Affairs*, 81(1), 31–52.

Buzan, Barry Ole Wæver and de Wilde, Jaap (1998) *Security: A New Framework for Analysis*. Boulder: Lynne Rienner.

Ceballos, Gerardo, Ehrlich, Paul R. and Dirzo, Rodolfo (2017) 'Biological annihilation via the ongoing sixth mass extinction signaled by vertebrate population losses and declines'. *Proceedings of the National Academy of Sciences*, 114(30), E6089–E6096.

Clark, Nigel, and Szerszynski, Bronislaw (2021) *Planetary Social Thought: The Anthropocene Challenge to the Social Sciences*. Cambridge: Polity Press.

CNA (Center for Naval Analysis) (2007) *National Security and the Threat of Climate Change*. Washington DC: Center for Naval Analysis.

CNA (2014) 'National security and the accelerating risks of climate change'. Available at: https://www.cna.org/cna_files/pdf/MAB_5-8-14.pdf

Crawford, Neta (2022) *The Pentagon, Climate Change, and War*. Cambridge, MA: MIT Press.

Crutzen, Paul J. (2016) The 'Anthropocene'. In Eckart Ehlers and Thomas Krafft (eds) *Earth System Science in the Anthropocene*. Berlin: Springer, pp 13–18.

Cudworth, Erika, and Hobden, Stephen (2017) 'Post-human security', in Anthony Burke and Rita Parker (eds) *Global Insecurity: Futures of Global Chaos and Governance*. London: Palgrave Macmillan, pp 65–81.

Dalby, Simon (2015) 'Climate change and the insecurity frame', in Shannon O'Lear and Simon Dalby (eds) *Reframing Climate Change: Constructing Ecological Geopolitics*. Abingdon: Routledge, pp 83–99.

Department of Defence, Australia (2023) *National Defence: Defence Strategic Review, 2023*. Canberra: Commonwealth of Australia.

Depledge, Duncan (2023) 'Low-carbon warfare: climate change, net zero and military operations'. *International Affairs*, 99(2), 667–685.

Deudney, Daniel (1990) 'The case against linking environmental degradation and national security'. *Millennium*, 19(3), 461–473.

Dunne, Tim (1998) *Inventing International Society: A History of the English School*. London: Palgrave.

Elhacham, Emily et al (2020) 'Global human-made mass exceeds all living biomass'. *Nature*, 588(7838), 442–444.

Evans, Jake (2024) 'Climate risks ignored in national defence strategy, former defence chief says', *ABC News*, 2 May.

Garcia, Denise (2010) 'Warming to a redefinition of international security: the consolidation of a norm concerning climate change'. *International Relations*, 24(2), 271–292.

Garcia-Gibson, Francisco (2023) 'The ethics of climate activism'. *WIRES Climate Change*, 14(4), e831.

Gardiner, Stephen (2013) *A Perfect Moral Storm: The Ethical Tragedy of Climate Change*. Oxford: Oxford University Press.

Grove, Jairus (2019) *Savage Ecology: War and Geopolitics at the End of the World*. Durham, NC: Duke University Press.

Hardt, Judith (2021) 'The United Nations Security Council at the forefront of (climate) change? Confusion, stalemate, ignorance'. *Politics and Governance*, 9(4), 5–15.

Hardt, Judith, Estève, Adrien, Harrington, Cameron, Simpson, Nicholas P. and von Lucke, Franziskus (eds) (2023) *Climate Security in the Anthropocene*. Berlin: Springer.

Hardt, Judith et al (2024) 'The challenges of the increasing institutionalization of climate security'. *PLOS Climate*, 3(4), e0000402.

Harrington, Cameron (2023) 'Climate change as a threat multiplier: the construction of climate security by the UK, 2007–2020', in Judith Hardt et al (eds) *Climate Security in the Anthropocene*. Berlin: Springer, pp 297–318.

Harrington, Cameron and Shearing, Clifford. (2017) *Security in the Anthropocene*. New York: Columbia University Press.

IPCC (Intergovernmental Panel on Climate Change) (2022) *Climate Change 2022: Impacts, Adaptation and Vulnerability*. Geneva: IPCC.

Levy, Marc A. (1995) 'Is the environment a national security issue?' *International Security*, 20(2), 35–62.

Litfin, Karen (1999) 'Constructing environmental security and ecological interdependence'. *Global Governance*, 5, 359–378.

Lövbrand, Eva et al (2015) 'Who speaks for the future of the Earth?' *Global Environmental Change*, 32, 211–218.

Maas, Achim, and Carius, Alexander (2012) 'Territorial integrity and sovereignty: climate change and security in the Pacific and beyond', in Jurgen Scheffren et al (eds) *Climate Change, Human Security and Violent Conflict*. Berlin: Springer, pp 651–665.

Maertens, Lucile (2021) 'Climatizing the UN Security Council'. *International Politics*, 58(4), 640–660.

McDonald, Matt (2013) 'Discourses of climate security'. *Political Geography*, 33, 42–51.

McDonald, Matt (2021) 'After the fires: climate change and security in Australia'. *Australian Journal of Political Science*, 56(1), 1–18.

McDonald, Matt (2023a) 'Geoengineering, climate change and ecological security'. *Environmental Politics*, 32(4), 565–585.

McDonald, Matt (2023b) 'Immovable objects? Impediments to a UN Security Council resolution on climate change'. *International Affairs*, 99(4), 1635–1651.

McDonald, Matt. (2024a) 'Accepting responsibility? Institutions and the security implications of climate change'. *Security Dialogue*, 55(3), 293–310.

McDonald, Matt (f2024b) 'Climate change, security and the institutional prospects for ecological security'. *Geoforum*, 155, 104096.

McNaught, Rebecca (2024) 'Visualising the invisible: collaborative approaches to local-level resilient development in the Pacific Islands region'. *Commonwealth Journal of Local Governance*, 26, 28–52.

Mische, Patricia (1989) 'Ecological security and the need to reconceptualize sovereignty'. *Alternatives*, 14(4), 389–427.

Mitchell, Audra (2014) 'Only human? A worldly approach to security'. *Security Dialogue*, 45(1), 5–21.

Mitra, Ryan, and Sanghi, Sanskriti (2023) 'The small island states in the Indo-Pacific: sovereignty lost?' *Asia Pacific Law Review*, 31(2), 428–450.

Mobjörk, Malin and Lövbrand, Eva (eds) (2021) *Anthropocene (In)Securities: Reflections on Collective Survival 50 Years after the Stockholm Conference*. Oxford: Oxford University Press.

Moore, Jason W. (2016) *Anthropocene or Capitalocene? Nature, History, and the Crisis of Capitalism*. San Francisco: PM Prbovaess.

Morgan, Wesley, Carter, George and Manoa, Fulori (2024) 'Leading from the frontline: a history of Pacific climate diplomacy'. *Journal of Pacific History*. https://doi.org/10.1080/00223344.2024.2360093

Myers, Norman (1993) *Ultimate Security: The Environmental Basis of Political Stability*. New York: W.W. Norton.

New Zealand Ministry of Defence (2019) *The Climate Crisis: Defence Readiness and Responsibilities*. Wellington: MoD.

PIF (Pacific Islands Forum) (2018) *Boe Declaration Action Plan*. Suva, Fiji: Pacific Islands Forum.

Pirages, Dennis and Cousins, Ken (eds) (2005) *From Resource Scarcity to Ecological Security*. Cambridge, MA: MIT Press.

Reed, Graeme et al (2021) 'Indigenous guardians as an emerging approach to indigenous environmental governance'. *Conservation Biology*, 35(1), 179–189.

Rockström, Johan et al (2009) 'A safe operating space for humanity'. *Nature*, 461(7263), 472–475.

Rockström, Johan et al (2023) 'Safe and just Earth system boundaries'. *Nature*, 619(7968), 102–111.

Scartozzi, Cesare M. (2022) 'Climate change in the UN Security Council: an analysis of discourses and organizational trends'. *International Studies Perspectives*, 23(3), 290–312.

Soper, Kate (1995) *What Is Nature? Culture, Politics and the Non-human*. Oxford: Blackwell.

Steffen, Will (2021) 'Introducing the Anthropocene: the human epoch'. *AMBIO*, 50(10), 1784–1787.

Tkachenko, Aly (2024) 'Envisioning ecological security through local direct action: community resistance with global resonance'. *Critical Studies on Security*. https://doi.org/10.1080/21624887.2024.2405275

Vogler, Anselm (2023) 'Tracking climate securitization'. *International Studies Review* 25(2): viad010.

Vogler, Anselm (2024) 'On (in-)secure grounds: how military forces interact with global environmental change'. *Journal of Global Security Studies*, 9(1), 1–19.

Wæver, Ole (1995) 'Securitization and desecuritization', in Ronnie D. Lipschutz (ed.) *On Security*. New York: Columbia University Press, pp 46–86.

Walt, Stephen (2021) 'The realist guide to solving climate change', *Foreign Policy*, 13 August.

Welsh, David A. (2022) *Security: A Philosophical Investigation.* Cambridge: Cambridge University Press.

Wendt, Alexander (1992) 'Anarchy is what states make of it: the social construction of power politics'. *International Organization*, 46(2), 391–425.

Williams, Michael C. (1998) 'Identity and the politics of security'. *European Journal of International Relations*, 4(2), 204–225.

Wilson Rowe, Elana Wilson (2021) 'Ecosystemic Politics: analyzing the consequences of speaking for adjacent nature on the global stage'. *Political Geography*, 91, 102497.

WWF (World Wide Fund for Nature) (2022) *Living Planet Report 2022.* Gland, Switzerland: WWF.

4

Nature's Hierarchies? Ecosystems and Order Making

Elana Wilson Rowe, Paul Beaumont and Lucas de Oliveira Paes

Introduction

Control of nature has long been intertwined with the exercise of political power. It was essential to the formation of the modern state and modes of 'mega-planning', and involved re-envisioning people and lands into simplified and standardized forms (Scott 2020). At the interstate level, studies have highlighted how control of – or ability to know nature in particular ways – has been an essential practice for reproducing civilizational hierarchies and the exercise of associated privileges. To capture this dynamic, scholars have explored topics from Antarctic governance and regulation of rivers in Europe (Yao 2021, 2022), through mastering the oceans (de Carvalho et al 2022) to participating in planetary governance (Lehman 2020). Developing this theme further, a growing strand of scholarship explores and illustrates how spatializations anchored in nature itself figure into the establishment of novel hierarchies, both globally and regionally (Wilson Rowe 2021; Beaumont and Wilson Rowe 2022; Paes 2022, 2023).

Picking up this line of analysis, this chapter examines how cooperation around border-crossing ecosystems can shape, generate and potentially transform international hierarchies. This inquiry is timely as the Anthropocene has given rise to intensified efforts to govern (or at least seem to govern) the world's ecosystems, yet such ecosystems do not respect sovereign borders. Indeed, hundreds traverse multiple states and thus require complex international cooperation. This chapter builds upon an emerging research agenda that critically examines the political and social consequences of anchoring cooperative arrangements in transboundary ecosystems. Specifically, we pick up on research that has called for exploring the broader ordering consequences of anchoring cooperation in border-crossing ecosystems and has shown how, across multiple instances, a central outcome is the emergence of durable and novel hierarchies of ecosystemic insiders and outsiders (Wilson Rowe 2021, forthcoming).

This chapter aims to further develop this insight on hierarchical change and ordering around ecosystems by considering three key cases – the Arctic, the Amazon and the Caspian Sea – and a broader suite of dynamics and expressions of ecosystem-anchored hierarchies. We suggest, by way of a conclusion, that one can identify not only the hierarchical outcomes that have been identified previously but also a process and sequence of ecosystem-anchored hierarchy formation. The chapter unfolds as follows. We start by exploring how ecosystems and the idea of ecosystems have been implicated in efforts to order political relations, both historically and today. Next, we further unpack the concept of hierarchy, making an argument that hierarchy is a useful lens through which to view changing power relations around emerging and established ecosystemic policy objects.

We then identify and compare three interrelated but analytically distinct processes through which ecosystem cooperation engenders hierarchical change via the Caspian Sea, Arctic, and Amazon cases. First, we show how Caspian Sea ecosystem cooperation brought together 'unlikely bedfellows' into a new political grouping and collective identity. Second, and relatedly, we look at the changing networks relating to the Arctic Council over time to highlight how ecosystem cooperation begets and legitimates sharper boundaries demarking 'ecosystemic insiders' from outsiders (non-ecosystem-adjacent states) and facilitate regional exclusionary dynamics. Third, based on the case of the Amazon, we show how boundary making around ecosystems can enable the inversion of global hierarchies, with regional gatekeeping empowering conventionally weaker actors.

The inquiry in this chapter connects to the broader questions explored in this volume about how the governance of nature can precipitate new measures and approaches to address environmental or ecosystem-specific challenges, but also catalyse the reordering of relations more generally. We argue that such a reorganization may happen regardless of whether the governance succeeds in its stated goals or results in improved management of natural, transboundary resources and ecosystems. Taken together, the chapter argues that heightened awareness of and pressure to govern through and across ecosystems can fundamentally shape and even transform the international order as we know it.

Ecosystems and ordering

The emergence of the idea of 'ecosystems' came about during a period in which the intellectual silos in natural knowledge, erected by Enlightenment science, were being challenged. While the Enlightenment period had hastened a growth in new forms of scientific knowledge, it involved dismantling and examining as separate elements: the botanical, biological, fluid, atmospheric and geological elements and forces that practitioner knowledge long appreciated and engaged with as a holistic and interconnected system (Turnbull 2005; Mills 2009; Cameron and Earley 2015; Roszko 2022). The need to reconceptualize nature (and its governance) related to some of the obvious political failures and shortcomings of 'high modernist' schemes that whisked away the complexities of nature – and social life – and prescribed overly narrow (and often highly utopian) solutions that failed their own ambitions and reproduced dominant social structures (Scott 2020).

Emergent understandings of ecosystems gave rise to new disciplines, like ecology and oceanography (Mills 2009, 2011), and were also implicated in new forms of political organization. For example, classifications of nature into particular, interconnected zones was fundamental to how the British imperial project in the Americas was structured and adapted to changing circumstances, from economics to security concerns (Greer 2015). Similarly, the cataloguing and portrayal of nature by early modern Spanish colonizers was a means of justifying domination of both resources and Indigenous peoples in Latin America (Caraccioli 2021). To take a more contemporary example, over 60 per cent of all international environmental agreements are not fully global, but rather agreed between some states (bilateral or more) (Balsiger and Prys 2016). Of the 2,227 international agreements analysed, these scholars find that a solid fourth of all agreements (589 agreements) match membership and spatial ambit – in other words, regionally tailored environmental cooperation designed to address issues in a particular area (Balsiger and Prys 2016, 249), while numerous environmental institutions also govern 'ecoregions' from the Andes to the Great Lakes of the US (Church 2020).

A novel dataset on ecosystems expands the inquiry beyond known examples in environmental politics to explore whether and how ecosystems are governed more generally (for environmental or other ends) (Maglia and Wilson Rowe 2023, 2024). This approach takes as its starting point that the anchoring of transboundary governance initiatives in ecosystems can generate enduring effects when it comes to the ordering of relations around an ecosystem (see Wilson Rowe 2021). This exploration of a world of large-scale ecosystems identified by natural science criteria has found that such meta-ecosystems are objects of governance in areas outside the environmental fields, and often below the level of formality or institution. Some ecosystems are governed through the member state composition of multi-issue regional multilateral bodies or have specific features of the ecosystem governed through issue-specific treaties (water rights, protection of a particular species, fisheries and so on). Furthermore, the dataset finds that some transboundary ecosystems with four or more adjacent states remain ungoverned, but actually surprisingly few (16 out of 105 terrestrial ecosystems, 7 out of 64 marine ecosystems or provinces and 2 out of 62 freshwater ecosystems).

Yet, the large-scale data gathering revealed that a significant subset of governed ecosystems are indeed governed through governance efforts that are established within and for the ecosystem itself. These cooperative efforts cover a range of issue areas, including social, economic or security-related issues, as well as some environmental elements we typically expect from cooperation around ecosystem. One can find this type of 'ecosystem-anchored' cooperation in 30 terrestrial systems, 17 marine ecosystems/marine provinces and 15 freshwater ecosystems. This specific kind of cooperation – generated with reference to natural interconnections and often drawing upon the technocratic language of environmental management – can become quickly naturalized and incontrovertible, as we will see in the cases of the Arctic, the Caspian Sea and the Amazon, which will be discussed later on. The ecosystems that 'emerge' as the focus of international cooperation have frequently undergone airbrushing, adjustments or outright reconfiguration. In the case of the Caspian Sea, for example, the tributary states highlighted as significant by scientists and international environmental organizations were removed from the picture entirely when political cooperation commenced (Beaumont and Wilson Rowe 2022). Similar in outcome if different in process, post-Cold War Arctic cooperation involved an amplification of a club identity among the Arctic states and Arctic actors, and a gradual marginalization of non-Arctic actors in the politics of the Arctic Council (see Wilson Rowe 2021, forthcoming).

The ways in which nature serves as a resource in global ordering is explored in depth by a strand of scholarship examining changing power relations around the globe's major border crossing ecosystems (Wilson Rowe 2018, 2021; Beaumont and Wilson Rowe 2022; Paes 2022b, 2023). This research has explored how drawing upon nature and ecosystems seems to serve as an effective resource in reordering relations among states. We argue that exploring these transformations through a hierarchy lens can illuminate changing power relations and both enable exploring change over time and facilitate comparison with other cases. We thereby make the case that nature and its spatialization merits further attention in IR's hierarchy studies agenda, particularly as the weight of the planetary and crises increasingly bears on global politics more broadly.

Hierarchy and nature

So, why do we turn to the question of hierarchy to consider how ecosystems and their governance impact on global ordering? Hierarchy is an object of increasing attention in international relations scholarship (Lake 2011; Mattern and Zarakol 2016; Zarakol 2017; MacDonald 2018; Mcconaughey, Musgrave and Nexon 2018; Musgrave and Nexon 2018; Beaumont 2024; Beaumont and Glaab 2023; Paes 2025). The focus on hierarchy has served as an important corrective to the field of International Relations, by directing attention to variegated ways in which global relations are ordered while insisting upon analytical sensibility and reflexivity towards how enduring power relations structure social outcomes. Nevertheless, the study of the ordering dynamics and power relations comprising hierarchy in world politics is far from homogeneous.

Zarakol (2017) distinguishes two branches scholarship on hierarchy: those who define hierarchy *narrowly* as an authority relationship and those who define it *broadly* as a structure of material and/or symbolic inequality. The narrow concept of hierarchy is seen as akin to a bargain, whether gently or roughly enforced, where states forgo their sovereign authority to other actors in exchange for the delivery of certain outcomes for all the actors involved (Lake 2007, 2009; MacDonald 2018). Alternatively, broadly defined hierarchies are expressions of more embedded forms of power in which actors' identities are so structured by longstanding forms of inequality that hierarchy becomes difficult to identify and, by extension, difficult to contest (Mattern and Zarakol 2016). Crucially for our purposes here, as one of the leading scholars of this field puts it, these 'deep hierarchies' demarcate those 'that belong, or do not, in some space of world politics' (Zarakol 2017: 7), by instantiating entrenched structures, power positions and relations. Notably, such broad hierarchies may cut across and/or envelop state boundaries such as hierarchies of race, gender and civilization (Sjoberg 2017; Suzuki 2017; Zarakol 2017; Yao 2019, 2021; Barder 2021).

In this sense, deep hierarchies work by naturalizing power positions and relations. However, until recently, this scholarship had not explored how representations of the natural world and attempts to govern it interplay with hierarchical positioning. Yet recent works have highlighted how discourses depicting the physical world can serve as a source of hierarchy: producing boundaries and asymmetric positionalities (Lewis and Wigen 1997; Gruby 2017; Depledge 2018; Wilson Rowe 2018, forthcoming; Beaumont and Wilson Rowe 2022; Paes 2022, 2023). As critical geopolitics has long explored, representations of space can facilitate, hide and ultimately naturalize relations of power. Here we show how that is also the case as states seek to delimit the scopes of ecosystemic governance.

Synthesizing the findings from the emergent ecosystem politics research agenda, the following analysis foregrounds and links the mechanisms of hierarchy formation to propose a processual account of how ecosystemic hierarchies emerge and develop. Indeed, zooming in on the emergence of regional institutions to govern the Arctic Ocean, the Amazon rainforest and the Caspian Sea, we specify and explore three dynamics pertaining to the emergence and evolution of hierarchical relations in our research on these cases and consider findings in a comparative perspective. In what follows, we consider changing regional relations as ecosystem cooperation emerges, in what ways hierarchies are enacted along often freshly drawn spatial boundaries of ecosystem-anchored cooperation and, finally, how these new hierarchies seem able to (sometimes) invert or at least subvert broader global hierarchies and power relations. As we will discuss in the conclusion, these dynamics can be seen as logically flowing and following from one another in a sequential but certainly not determinative fashion.

Unusual bedfellows? Rethinking relations through nature in the Caspian Sea

The Caspian Sea region's strategic importance as transit route and the discovery of latent riches beneath the seabed provided the stakes for what (especially Western) observers long worried was a regional tinderbox (see Bayramov 2021). Indeed, the Caspian states – Azerbaijan, Iran, Kazakhstan, Turkmenistan and Russia – had long struggled to allocate rights over the resources: a two-decade-long quarrel prevented legal partition and has been frequently punctuated with hostilities that bordered upon war (see Bayramov 2021). Yet, these states decided to begin a comprehensive cooperation around the Caspian Sea as a fragile – and wealth-producing – ecosystem after the dissolution of the Soviet Union disrupted longstanding bilateral Soviet-Iranian agreements. This early environmental cooperation gave impetus to resolving broader issues and, at the 2018 Aktau summit, the leaders of the five Caspian states signed an agreement that went a long way to resolving longstanding legal disputes on the status and division of authority in the Caspian Sea.

As we will discuss later on, the political boundaries of the Caspian Sea ecosystem have not been a policy given, but have developed and solidified out of efforts to cooperate regionally. The collapse of the Soviet Union led to a lapse in the bilateral (with Iran) treaties that had governed some of the border-crossing interests associated with the Caspian Sea, such as fisheries and access of nonregional actors to the Sea. In enacting a period of new cooperation, international environmental organizations played a crucial role (Bayramov 2020, 2022). Notably, they initially pitched a much wider conception of the Caspian Sea than the main body of water and its adjacent states: on the basis of expert input, they initially recommended that cooperation include the states with important tributaries leading into the sea (the Caspian basin), which was soon rejected for being politically impractical (Beaumont and Wilson Rowe 2022). Among the states themselves, there was disagreement about whether the Caspian should be considered a sea or a large inland lake, along with the different forms of international legal guidance and strictures that such a definition would proffer. In 2018, the states concluded a legal agreement on the status of the Caspian Sea that resolved many of these issues and became the legal cornerstone of the emergent 'Caspian 5' collective identity – the five adjacent littoral states.

The emergence of ecosystemic politics around the Caspian Sea highlights the first hierarchical dynamics we identify as born out of efforts to govern cross-border ecosystems; bringing together 'unlikely bedfellows' into a new political grouping. In our cases, sharing this common political project grounded in ecosystem adjacency appears to offer an opportunity to innovate with relations, and insulate, overcome or at least legitimate overlooking pre-existing rivalries and animosity. The Caspian Sea, in particular, offers an illustration of how ecosystem discourse-engendered cooperation can provide a scientifically legitimate rationale for open-ended and thoroughgoing cooperation between unlikely bedfellows. Indeed, prior to the Tehran Convention, all the Caspian states had never been members of the same international organization, let alone one exclusive to what these states themselves now call the 'Caspian Five'. Moreover, the early moves that would lead the way to the Tehran Convention would occur in parallel with growing contention between littoral states regarding whether the water that divided them was a sea, a lake or a condominium, and thus how to divide its resources. In other words, ecosystem cooperation brought together states that otherwise not only did not cooperate well in other contexts, but were frequently at odds and openly hostile.

As Bayramov (2022) has argued, the timing and intensity of cooperation among the Caspian states suggests that environmental cooperation led or fed into cooperation on security (2010) and then the legal convention (2018). Indeed, Bayramov (2022) makes a neofunctionalist argument that the technical and relatively low salience of the character of environmental cooperation-built trust and developed cooperative habits, facilitating negotiations in more contentious areas. Three features of the cooperative processes that led to the Tehran Convention and ecosystem cooperation are worth emphasizing: (1) it represented the first sustained cooperation among the Caspian states; (2) it saw the Caspian ecosystem becoming a distinct object of global governance and international intervention; and (3) it knitted together previously uncooperative actors in a politically awkward space (Beaumont and Wilson Rowe 2022).

At first blush, a sceptic might point out that cooperation between littoral states around a common resource is logical and the fact that the Caspian Sea is conceived by those states as an ecosystem is epiphenomenal to the outcome. While sceptics might concede that the states were not natural bedfellows, they may point out that environmental peacebuilding research has long highlighted how the relatively low political stakes in environmental cooperation can enable rivals and even enemies to cooperate and generate spillover effects into higher political domains. This is precisely Bayramov's (2022) argument: that IOs catalysed environmental cooperation and thus smoothed the path to the 2018 Convention. However, we contend that the specific qualities of the ecosystem discourse (vis-à-vis other kinds of environmental cooperation) and the associated practices that grew out of the Tehran Convention have been consequential beyond providing a positive environment for tackling challenging political issues.

Indeed, the Caspian Sea cooperation endorsed ecosystem management, which required envisioning a deeper form of cooperation than would have been the case with issue-specific environmental cooperation, species-conservation or traditional fish stock management. Here, a shared view of ecosystem interdependence was crucial for welding the states' economic interests to a broad set of environmental cooperative activities. The intention in the planning documents was quite explicitly to 'link biodiversity conservation and fishery production objectives to advance EBM [ecosystem-based management] in the Caspian Sea' (UNDP 2008: 26). This theory of ecosystem interdependence enabled issues and problems that might once have been treated in isolation to become irreducibly connected to the health of the ecosystem whole: 'The underlying and root causes of unsustainable bioresource utilization ... were poor regional management, overfishing, illegal fishing and pollution remain valid but the productivity and integrity of the ecosystem is also now recognized as an underlying cause.' Thus, conceptualizing the Caspian Sea's environmental problem in terms of ecosystem health served to legitimate both the geographical scale of the cooperation and its cross-cutting scope. As one of the planning documents warned, unless the Caspian states adopt a regional ecosystem management approach, they will struggle to 'integrate fishery management and biodiversity conservation objectives' (UNDP 2008: 55). A corollary to this is that *comprehensive* regional cooperation becomes the only scientifically legitimate means of solving the Caspian Sea's environmental problems. The Tehran Convention (2006) that resulted from this process reflected EBM's logic: it established a framework for Caspian cooperation that would be developed through *six* further additional protocols, which would be steadily negotiated over the course of the next decade.

While perhaps most salient in cooperation regarding the Caspian Sea, cooperation among countries with limited track records of cooperative engagement and even previous rivalries is also present in the Arctic and the Amazon. In the latter case, the Amazon region before the emergence of ecosystem-based cooperation was a point of sparser cooperation amid the other cooperation initiatives that had been created in Latin America in the previous years. The Amazon Cooperation Treaty (ACT) brought together Caribbean and Andean countries, which had already initiated bilateral cooperation with Brazil, whose integration had also been then mostly focused on its southern non-Amazon neighbours (Paes 2022). In the Arctic, cooperation regarding the Arctic ecosystem was also one of the few elements of cooperation that transected the chilly consequences of the Cold War, and active, ecosystem-based cooperation in the post-Cold War period was meant to actively erase former Cold War divisions (English 2013; Steinberg et al 2015) . Taken together, all three cases highlight how ecosystem cooperation can, given the divergence between ecosystems and state borders, *likely* involve the drawing of new spatialized boundaries and bringing together actors that have not cooperated extensively previously.

Enhancing boundaries: creating ecosystem insiders and outsiders in the governance of the Arctic

Although most of us could point to the Arctic on a globe, the Arctic policy field is wide and varied and maps onto Arctic ecosystems with a variety of governance measures, from bilateral arrangements in the Bering and Barents Seas through proposed Indigenous-led governance of the Arctic waters connecting an Inuit homeland to the eight countries that convene alongside Arctic Indigenous peoples to discuss issues of an Arctic meta-ecosystem in the Arctic Council (Dodds and Nuttall 2016; Pikialasorsuaq Commission 2017; Burke 2019; Rottem 2020; Wood-Donnelly 2020). Where or what the Arctic ecosystem or region consists of is the subject of competing and overlapping definitions. Some would point to the slightly varying extent of the tree line, a squiggly line across the circumference of the globe indicating the emergence of conditions beyond which trees will not grow. Others use the lines of latitude, with the Arctic Circle, which marks the southernmost latitude at which – during the peak of winter – the sun will not rise all day. The 'Arctic' is, in sum, a vastly varied physical space of 'many Arctics' – from the sea ice-bound and roadless Canadian North to Russia's large swath of permafrost to the comparatively temperate Nordic Arctic. As we can see in Figure 4.1, the Arctic Ocean itself is interconnected and is also an assemblage of smaller seas subject to varied freshwater inland flows, current patterns and interactions with the wider world (in the case of Figure 4.2, streams of plastic pollution). Despite this physical diversity, for our purposes, the longstanding policy emphasis on the Arctic as an interlinked ecosystem renders it appropriate for considering how ecosystemic hierarchies are established, a question to which we now turn.

This circumpolar ecosystem-anchored cooperation primarily takes place in the Arctic Council and in associated efforts at binding regional treaties, including search and rescue, scientific cooperation, and oil spill preparedness and response. The eight-country Arctic Council is the highest-level multilateral setting for Arctic issues with the most comprehensive membership of all Arctic/northern forums and has undertaken two decades

Figure 4.1: Map of protected areas of the Caspian Sea showing river tributaries and nearby countries

Source: GRID-Arendal 2012

of extensive cooperative work on both science and policy responses to Arctic governance challenges. As with the Caspian Sea cooperation, how the natural delimitations to the Arctic ecosystem(s) are drawn is a political project. In other words, where or what the Arctic consists of is the subject of competing and overlapping definitions, and allows for or privileges the participation of certain actors (English 2013; Shadian 2014; Steinberg et al

Figure 4.2: Map of plastic input into the Arctic Ocean

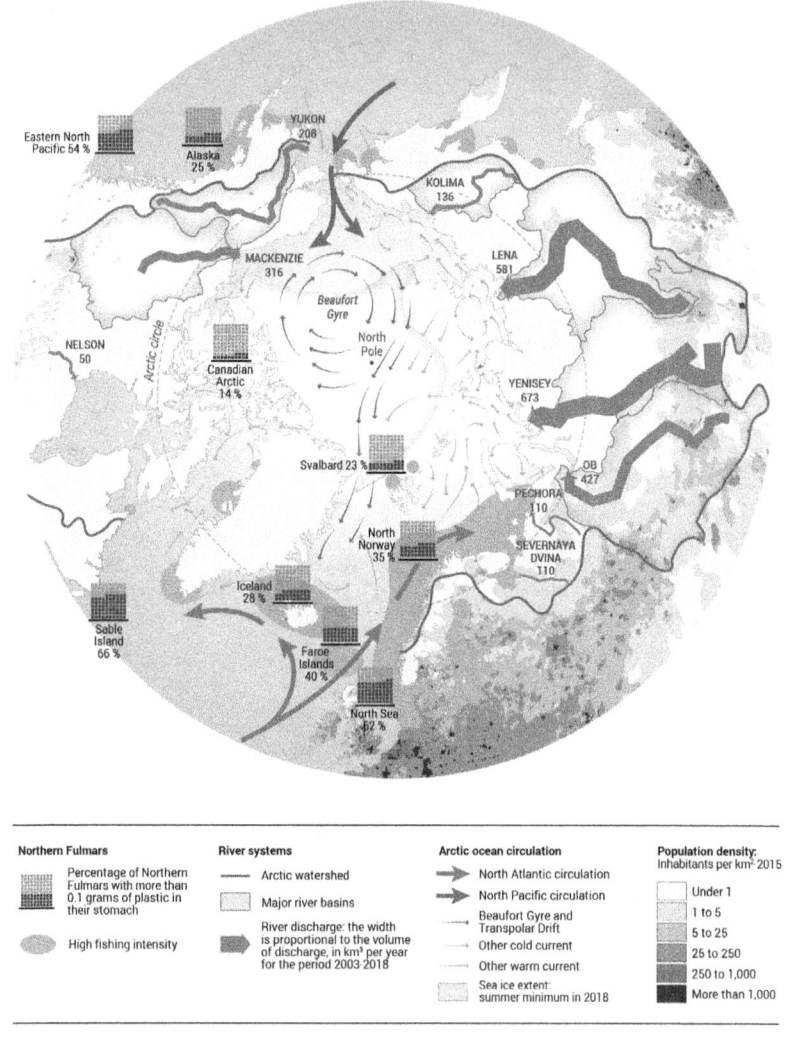

Source: Rekacewicz et al 2019

2015; Dodds and Nuttall 2016; Raspotnik 2018; Wilson Rowe 2018; Gad and Strandsbjerg 2019).

In this sense, the governance of the Arctic illustrates how cooperation on ecosystems offers an opportunity for the creation of boundaries between insiders and outsiders, in which actors mutually perceived as being operating 'within' the ecosystem are distinguished from those cast as external to it. We observe here that these insider actors progressively construct regional cooperation, sustaining and reinforcing their privilege and legitimacy as the ultimate governors of their shared ecosystem, and gatekeep access to those deemed as outsiders. This spatial boundary demarcation thus produces a hierarchy in terms of which actors are entitled to influence the ecosystem's governance (see Wilson Rowe [2021] for an exploration of this dynamic). The clustering of 'unusual bedfellows' into the 'Caspian states' as previously mentioned was far from a neutral or obvious outcome. Rather, it was a political outcome and process that involved excluding other countries (tributary states) and diminishing the role of other actors (specifically the international organizations that had initiated much of the cooperation relating to the Caspian Sea). Similarly, in the case of post-Cold War Arctic cooperation, there have been clear, repeated investments of diplomatic energy and capital in the maintenance of an Arctic-level geopolitical framing of northern challenges. A growing scholarship has illustrated, through a variety of methods, how the anchoring of Arctic governance so squarely– with ecosystem adjacency as a central form of political capital – may have been one of the most significant political outcomes of post-Cold War Arctic cooperation (see Dodds and Nuttall [2016] and Depledge [2018] on polarization and circumpolarization more generally).

The political significance of this bounding of the ecosystem is evidenced by how frequently the lines between regional and global politics have been revisited and enforced in the Arctic Council (Wilson Rowe 2018). On the other hand, all working groups now have international or global interfaces with other actors and organizations. Also, a growing set of non-Arctic state and civil society observers as well as the most recent chairmanship platform highlighted the need to work more proactively within international fora, for example on marine plastic pollution. In this light, one could read the occasional shoring up of the Arctic political boundary as an occasional political feat, responding to certain issues and realizing preferred power relations for particular Arctic moments, rather than a more durable ordering of global relations.

However, if we take a longer-term view and examine the changing actor picture over time, the ways in which an ecosystem can be entangled with and serve to structure global politics becomes clearer. A study exploring the position of Arctic and non-Arctic actors in Arctic Council policy making over three decades finds evidence for a strengthening norm over time to diminish discussion time devoted to nonregional actors and international organizations (Wilson Rowe 2021). By contrast, in the early stages of this cooperation around the Arctic ecosystem, there are frequent and positive references about the significance of potential contributions that could be made to Arctic cooperation by the Global Environmental Facility (GEF), United Nations Environment Programme and the EU, among others (Wilson Rowe 2021). Historical studies of the early days of Arctic Council development have noted a similar starting point, with the involvement of nonstate and non-Arctic actors seen by many states as unproblematic or natural given the nature of the Arctic ecosystem challenges that were to be addressed (English 2013).

The overall number of references – positive or negative – to international institutions and non-Arctic actors and initiatives in Arctic Council meeting reports declined precipitously as Arctic Council cooperation matured, with mentions plummeting from 54 in 1998–2000 to 25 in 2017–2019 (Wilson Rowe 2021: 7). In other words, the more organized Arctic cooperation became, the less necessary or less acceptable it became to discuss non-Arctic/globally anchored initiatives of relevance.

A similar dynamic between Arctic and non-Arctic actors can be discerned when it comes to who intervenes in ways to shape the progress of politics or the actions of others in the Arctic policy field. If we think of interventions as indicative of a diplomatic stature, making a larger number of statements (and having these recorded in the official minutes) would suggest that the others viewed the speaking actor as having a greater relevance or as having a greater legitimate demand on the time of those gathered than other actors. We have taken this to be an initial indication of leadership dynamics and a hierarchy among Arctic actors. Similar to the decline in number of times non-Arctic based actors are mentioned at all, we also see that they also intervene (or are given the opportunity to intervene) decreasingly over time, from 9 per cent of total statements made in Arctic Council minutes in 1998–2000 to 4 per cent in 2017–2019 (Wilson Rowe 2021: 7). Interestingly, these indicators also show the growing stature of nonstate but Arctic-based actors (these include Indigenous peoples, working groups and Arctic-based nongovernmental organizations [NGOs]). In this way, we see how a key source of political capital in this cooperative setting is Arctic adjacency/location for a range of actors, including both state and nonstate.

Is this same preference for Arctic state leadership manifested in how central these actors are within the broader networks around the Arctic Council? A look at relationships between actors and how this has changed over time suggests that non-Arctic actors have moved to more marginal positions in the policy networks relating to the Council. One key finding of the network analysis is that a relatively small number of actors have active relationships with other actors in the network, despite ever-expanding lists of participants and observers at Arctic Council meetings. In fact, the density of the network around the Arctic Council has decreased significantly over time (from 0.074 in the American chairmanship to 0.3 in the later Norwegian and Finnish chairmanships). Density is a measure of the actual number of relationships compared to the number of theoretically possible relationships between actors in a network. The decreasing density score shows that as the Arctic Council has deepened its cooperation and increased its activities, the relational activity has grown more focused on a comparatively smaller number of actors. If we look at the positioning of individual actors (nodes) and their relations, a similar trend to the diplomatic stature and 'speaking time' analysis mentioned earlier emerges. While eight non-Arctic organizations and actors had high degree centrality measures in the early days of Arctic Council cooperation (1998–2000), indicating that they maintained extensive links to many other actors within the network, the EU and the WWF the only non-Arctic actors still central in the Arctic network by 2017–2019 (Wilson Rowe 2021: 7).

Three decades of cooperation around the Arctic ecosystem has resulted in countless reports and studies of best practices, shaped our understanding of how climate change impacts the peoples and environment of the region, and forged a new diplomatic norm around the inclusion of rights holders (Indigenous peoples). This cooperation has also forged a durable structuring of global politics around the Arctic as a policy object, with a clear privileging of ecosystem-adjacent actors (state or otherwise) and the effective subordination of states and actors further afield. The shock delivered to the cooperation by Russia's reinvasion of Ukraine was thorough and work in the Arctic Council has been significantly complicated by the necessity of issuing a clear diplomatic response. At the same time, at earliest possible moment, albeit in a truncated fashion, Arctic states sought to revive cooperation in the Arctic Council, including Russia. Russia, when complaining about the lack of full-fledged pre-invasion style cooperation in the Arctic Council, frequently referred to China and the possibility of deepening its cooperation with non-Arctic states should cooperation with the Arctic states stall indefinitely (Wilson Rowe forthcoming). The signalling of and toying with this boundary suggests that anchoring cooperation in the Arctic ecosystem is also seen by the Arctic states themselves as a political resource and an important outcome of the post-Cold War cooperation.

Ultimately, their cooperation has resulted in an enduring hierarchy of ecosystem 'insiders' and ecosystem 'outsiders' that serves to pattern behaviour both within the Arctic and outside of it. This pattern is not limited to the Arctic cooperation. Beaumont and Wilson Rowe (2023) identify a similar solidification and sharpening of the social and political boundaries between ecosystem insiders and outsiders that goes beyond the technocratic requirements of cooperation. For instance, the Caspian states collectively celebrate Caspian Day to honour the Tehran Convention, and have self-consciously nurtured 'Caspian 5' identity in which ecosystem cooperation and collective conservation of the sea serves as a totem of the states' self-proclaimed 'good neighbourliness'. The ACT was also born out of an attempt of shielding Amazon-adjacent states from perceived threats of external influence and would evolve as a tool for retaining control of which global actors could participate in the governance of the ecosystem (Paes 2022, 2023). While the in-groups in both cases are yet concatenate into global actor with agency beyond the region, our final Amazon case illustrates this possible outcome of ecosystemic cooperation.

Inverting global hierarchies? Amazon regionalism and the strengthening of Global South voices in environmental governance

The images of a green tropical vastness crisscrossed by a web of sinuous rivers easily come to mind when one mentions the Amazon. However, as with the Arctic and the Caspian, the boundaries of the Amazon ecosystem are subject to multiple definitions. While few would dispute defining the Amazon ecosystem as the complex biodiversity springing from the river basin carrying the same name, its outer edges can be defined by various hydrographical and biotic criteria (Tigre 2017). The mapping of such biological demarcations onto political and administrative borders renders the definition of where the Amazon starts and ends further malleable and political. The Amazon ecosystem is politically divided into multiple national and subnational borders, indigenous peoples' territories and protected areas (Figure 4.3). No matter how measured, the sheer dimension of the Amazon is indisputable: the Amazon Basin drainage ecosystem occupies almost 40 per cent of South America and represents 56 per cent of all 'broad leaf forests' globally (Barbosa 2017: 1). However, the greatness of the Amazon is also not immutable. As the map in Figure 4.3 also shows, the boundaries of the Amazon are not only blurry by also moving. Anthropic uses and associated degradation (in darker shades) have shrunk the biotic frontier of the Amazon's forest biodiversity kilometres short of the outer scope of the basin and its administrative frontiers.

Despite all the blurry and moving boundaries that the Amazon ecosystem affords, in international cooperation, there is one boundary that trumps all others: that of the states that can claim sovereignty over that ecosystem. Preserving the line that sharply divides adjacent from non-adjacent states has been a central force driving the emergence and evolution of regionalism in the Amazon (Paes 2022, 2023). In 1978, by signing the ACT, Amazonian states ushered in a focal point for preserving and reproducing that boundary.

Before the ACT, the Amazonian states had a history of limited collective engagement. For centuries, cooperation (and conflict) was bilateral and connected to border disputes (though often in the Amazon) (Little 2001; Garcia 2011). For states with sovereignty over this ecosystem, the Amazon has long been the frontier of their political authority, 'a space to be territorialized' (Paes 2023: 64). However, it was the recognition of the Amazon as a cohesive natural unit, as the tropical biome springing from the Amazon River basin, that would produce regional cooperation. The emergence of the ACT in the 1970s and its evolution into an international organization in the 1990s are intimately tied to the parallel evolution of global environmental governance

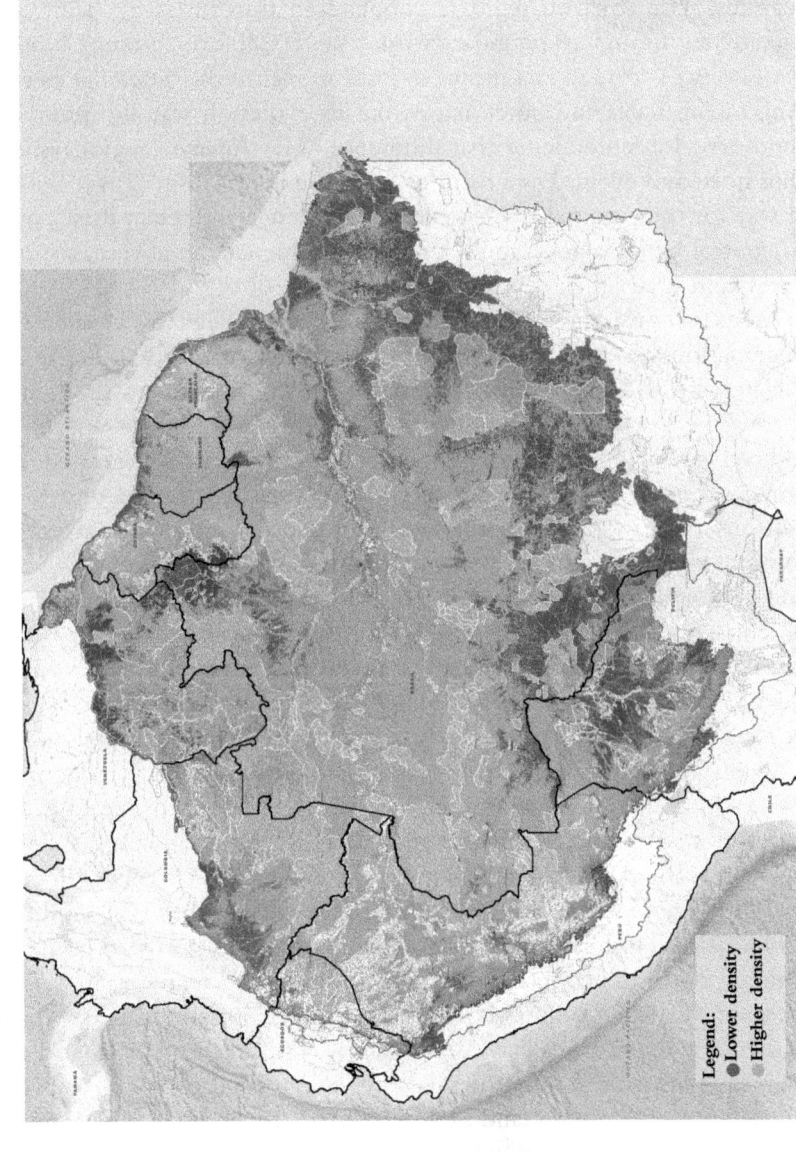

Figure 4.3: Amazon ecosystem boundaries and forest carbon density

Source: RAISG 2022

(Paes forthcoming). As Paes (2022, 2023) has demonstrated, the emergence of the regionalism in the Amazon was a to greater extent a reaction to global concern with the ecosystem's degradation and with the sustainability of the services it provides. Amazon regionalism would provide a platform for adjacent states to enhance their influence over the global environmental regimes. The Amazon regional cooperation thus interplays with global hierarchies through its boundary work (Paes 2023, forthcoming). Not only was the ACT created as a means to institutionalize the difference between Amazonian states and other actors, but its evolution was also profoundly motivated by reproducing that difference. The Amazon regionalism was thus born and evolved as a form of defensive regionalism (Tussie 2009) or 'a way for these states to block external influence and retain their capacity to control which actors would be able to influence the governance of the ecosystem under their sovereignty' (Paes 2023: 66).

At its origins, the ACT was a process mostly led by Brazil, where degradation was advancing most rapidly and where the ruling military dictatorship had long cultivated fears of foreign interests in the riches of the region (Becker 1982; Bratman 2019). In fact, fears of foreign economic intervention disguised as a legitimate interest in scientific knowledge and conservation had been already key to halting the creation of a United Nations Educational, Scientific and Cultural Organization (UNESCO) institute in the region in the 1940s. Growing global environmental awareness and conservationist concerns in international politics were thus seen as a potential threat to national security. In this sense, from the very outset, Amazon regional cooperation was an exercise in boundary making to shield external influence. Yet, the legitimation of such a boundary in terms of territorial sovereignty entailed an effort by the Brazilian diplomacy to downplay its own power asymmetry vis-à-vis its Amazon neighbours and assuage concerns of hegemonic ambition. Negotiations for the ACT came just as four Amazonian states (Bolivia, Colombia, Ecuador and Peru) had signed their own regional integration organization, the Andean Community, in 1969. These states, along with Venezuela, feared the ACT could be a tool for Brazil to hijack regional cooperation with its northern neighbours (Nunes 2016). The Brazilian diplomacy mitigated fears by limiting the scope of the treaty to the issues of cross-border environmental governance, particularly water, fauna and flora management, and removing any plans for political, economic or infrastructural integration (Landau 1981). Focusing on the defence of common (and exclusive) authority over a natural ecosystem as their point of convergence then enabled Amazon states to construct governance mechanism that flattened power asymmetries and distrust that would have precluded regional cooperation.

Through the first decades of the ACT, evolving local and global concerns with the environmental degradation of the Amazon and its implications for local Indigenous peoples and settler populations would raise the transnational pressure on the Amazon states. Amazon states responded to this growing pressure with growing coordination. The coordination was primarily aimed at achieving collective action to preserve their joint role as the sole representatives and gatekeepers of the region. This strategy found resonance in the notions of sustainable development and 'common but differentiated responsibilities', with their growing relevance in the aftermath of the Brundtland Commission and in the preparations of the United Nations Conference on Environment and Development (UNCED) in Rio de Janeiro in 1992 (Bratman 2019). The Amazonian states would subscribe to the idea of being stewards of international demands for sustainable development and demand financial assistance from the interested international stakeholders (ACTO 2013: 36–43).

The move of Amazon states from resisting the construction global environmental regimes to actively seeking to shape them would enhance their voices in global governance. Amazon cooperation would then gain strength as a channel for global-regional interplay. The ACT was converted into a formal organization – the Amazon Cooperation Treaty Organization (ACTO) in 2002, endowed with a permanent secretariat and helped coordinate the Amazon states' efforts to secure their centrality as the main governors of the region.

In this sense, the ACTO bolstered the ability of Amazonian states to act collectively to gatekeep external interest and funding, as well as their ability to retain the main representation of the ecosystem vis-à-vis external actors (Paes 2022). While being able to attract funding and activity, the ACTO still defers to member states regarding the destination of its resources (Garcia 2011; Pereira and Viola 2019, 2020). Additionally, its structure subordinates the presence of local nonstate actors to the national state's control. Therefore, Amazon regional cooperation produced a club of states that retain their position as the main stakeholders in the governance of the Amazon both regionally and globally.

Regional cooperation in the Amazon thus plays very interestingly with global hierarchies. It institutionalized the boundary afforded by ecosystem adjacency and the norm of permanent sovereignty over natural resources, allowing Amazon states to relegate global and transnational stakeholders to a peripheral position in regional governance. In this sense, conventionally powerful actors such as key donors and international organizations who often control dynamics of legitimation in global governance could be subordinated.

At the same time, the very emergence of the Amazonian regionalism is born at a historical moment when key hierarchical structures of modernity such as the notions of linear progress, development and civilization begin to face contestation for their environmental implications. The ACT was created as part of an effort by its members to protect their economic exploitation of the Amazon ecosystem, and to subordinate nature to the standards of civilization upon which international order was built. This was also at the core of the disregard that these development projects had for the ways of living predominant among Indigenous populations. The growing awareness of the impacts that the modes of economic progress driving modern societies have over the natural environmental was in itself a reversal of such hierarchies which Amazonian states initially resisted.

This conflictual logic of hierarchy, where markers of status associated with economic progress and environmental stewardship clash, have been consistently present in the politics of the Amazon to date. While the view of the Amazon as a pool of natural resources to the service of economic gains has been strongly contested within Amazon states themselves, it is far from extinct (as Bolsonaro and its ample support in parts the Amazon region remind us) (Viola and Franchini 2017; Pereira and Viola 2019). At the same time, the hierarchies of civilization based on stewardship have often also mobilized adjacency and scale as means for legitimation. Local forms of knowledge by Indigenous populations and sustainable extractive communities are key components for the legitimation of governance initiatives in the region.

The Amazon case therefore illustrates how cooperation on ecosystems can facilitate the inversion of structural hierarchies permeating world politics, raising the authority of conventionally peripheral actors. While far less pronounced and developed than in the Amazon, the Caspian five discourse and practice does engender representations that cut against global hierarchies. As numerous scholars have pointed out, the Caspian states (and the surrounding Central East Asian states) have long been orientalized and stereotyped in what Heathershaw and Megoran (2011) call a 'discourse of danger' in Western policy circles. In light of this conventional wisdom, we argue that the Caspian Five's emphasis on good neighbourliness and state capacity are not only descriptions of a functional ecosystem segment, but also claims to a social position that counters the prevailing and mostly negative narratives. Moreover, it may – and in this way, we echo but invert Yao's (2019, 2021) argument linking nature and hierarchy – enable the Caspian states to fend off unwanted international interventions and actors. Although not yet rivalling Amazon and Arctic states' efforts to club together against global interventions (Wilson Rowe 2021; Paes 2022b), Caspian environmental cooperation has already been used to explicitly reject international organizations' and non-Caspian states' efforts to address declining sturgeon stocks within international fora. In this sense, the inversion of global hierarchies seems to be more salient in the Global South cases, the Amazon rainforest and the Caspian Sea, where local states mobilize their sovereign authority to negotiate the influence of Global North actors. While Arctic states have yet to speak with one voice when it comes to global institutional settings, cooperating together robustly in and around the Arctic Council has served to order relations with extra-regional actors. Indeed, this logic of hierarchy inversion is reflected in how several local nonstate actors, often marginalized in both domestic and global politics, gain a stronger voice than non-adjacent states and international organizations.

Conclusion

This chapter has shown how cooperation on ecosystems has generated similar hierarchical effects across three very different cases. Indeed, we saw how in sites as different as the Arctic, Amazon and Caspian, ecosystem cooperation engendered three analytically distinct dynamics that can usefully be understood in terms of the formation and solidification of broad hierarchies. As each case illustrated, these broad hierarchies can shape and even transform the ordering of states in the region and the region's political and social relationship with the rest of the world. The fact that similar dynamics unfold in such different regional contexts hence lends weight to Wilson Rowe's (forthcoming) argument that cooperation around major, transboundary ecosystems can generate particular ordering effects. However, synthesizing across the three cases, and developing this line of argument further, we theorize that these processes of ecosystem-hierarchy formation are logically prone to occur in a *sequence* made up of three steps.

First, the identification of the ecosystem and the application of ecosystem scientific discourse provide a scientific rationale for cooperating with unusual partners. The substance of ecosystem management discourse requires the cooperation of all adjacent states, while the interconnectedness of ecosystems generates the need for complex and thicker cooperation than single-issue environmental discourse. Hence, ecosystem cooperation provides both new boundaries and an endogenous mechanism for close multilateral cooperation. Further research could utilize a politics of knowledge lens to explore how the emergence of such ecosystem policy objects relates to the privileging of certain forms of expertise in global governance (Esguerra 2024), and how the definition of an ecosystem policy object facilitates certain forms of governance and intervention (see Corry [2024] on weak and strong climate governance).

Second, once initiated, ecosystem cooperation appears to lend itself to practices that redraw and reify the boundaries between insiders and outsiders, such that external actors are systematically – consciously or not – excluded over time from intervening in ecosystem cooperation. The mechanism here may be generic to in-group cooperation over complex issues, but nonetheless, the specific boundaries, the endless time horizon built into ecosystem management logic and the substance of the cooperation gives this effect a distinct ecosyste*mic* flavour.

Third, the most developed form of ecosystem hierarchical effect – primarily found in the Amazon case but also detectable among the Caspian Sea – is that ecosystem cooperation may enable the states involved to invert or subvert longstanding global hierarchies and successfully claim authority vis-à-vis outsiders. The mechanism rests upon the well-established norm and value environmental stewardship, which ecosystem states can latch on to and deploy to rebut international actors and valorize the ecosystem-adjacent states. At its most advanced, the ecosystemic binding and bounding processes may generate a new global actor capable of exerting influence in domains beyond the ecosystem and the region.

These three steps should be read as an ideal-typical account of how ecosystemic politics *can* unfold. As our Arctic case illustrates, all three steps may not proceed exactly as theorized. Indeed, a fruitful avenue of research could be to explore when and why the ecosystem logic does not follow this pattern. Why might the identification of a transboundary ecosystem not generate cooperation among unusual bedfellows: are some animosities too powerful for ecosystem logic to overcome? Why has the deep and long-running Arctic cooperation not led to the inversion of global hierarchies or actor-hood beyond the Arctic? Do the characteristics of the states involved – their place in global hierarchies – influence the unfolding of ecosystemic effects?

If our sequential model of ecosystem-hierarchy formation is correct, then answering these questions would promise to shed light on the future configuration of regional and global orders in the coming decades. Indeed, with the heightened – unavoidable – recognition of the material and political transformation ushered in with the Anthropocene the global push for ecosystem cooperation is almost certainly likely to increase. Thus, if we are right, ecosystemic political effects are only likely to become stronger and more salient too.

Put into the context of this volume, we have shown that ecosystemic politics constitutes one viable means to explore how the practice of governing nature can feedback and even transform global order. Indeed, to varying extents, ecosystem politics in each of our cases generated new objects of global governance, shaped and reshuffled the regional actors populating world politics and instigated processes through which these collectives become solidified. These are all processes that would remain unseen or only be noted in passing in works that focus on evaluating and analysing whether and to what extent regional environmental institutions in meeting their stated goals. This line of reasoning begs two salient questions for understanding 'governing nature and global ordering'. First, does the governing of other 'natural' phenomena generate systematic effects? One long-term research agenda that could eventually emerge would be to compare efforts to govern transboundary ecosystems with politics generated around other natural phenomena. For instance, international relations scholars are starting to look at how tropical storms and the global response to them can transform the international political economy of the countries involved (Michelsen nd). In a broader sense, a comparative agenda could also be pursued by bringing ecosystems and their governance into the burgeoning research agenda on 'object-centred international relations', which seeks to make sense of the contours of global governance by taking a starting point in the objects of governance (for example, human rights, the climate and cyberspace) and the systems of expertise that define and delineate them, rather than privileging actors and their interests (Corry 2024; Esguerra 2024). Second, circling back to the core concern of global environmental politics, how do these broader consequences, engendered by the governing of nature, feed back into the effectiveness of governing nature? Indeed, it remains an open question in each of the case studies given in this chapter as to whether the solidification of insiders and outsiders helps or hinders environmental stewardship. At a minimum, identifying these processes should prove useful for identifying lurking variables that rational-intuitionalist approaches are prone to overlook.

References

Amazon Cooperation Treaty Organization (ACTO). 2013. *Legal Basis of the Amazon Cooperation Treaty.*

Balsiger, Jörg and Miriam Prys. 2016. 'Regional Agreements in International Environmental Politics'. *International Environmental Agreements: Politics, Law and Economics* 16(2): 239–260.

Barbosa, Luiz C. 2017. *Guardians of the Brazilian Amazon Rainforest: Environmental Organizations and Development.* Abingdon: Routledge.

Barder, Alexander D. 2021. *Global Race War: International Politics and Racial Hierarchy.* Oxford: Oxford University Press.

Bayramov, A. 2020. 'The Reality of Environmental Cooperation and the Convention on the Legal Status of the Caspian Sea'. *Central Asian Survey*, 39(4): 500–519.

Bayramov, A. 2021. 'Conflict, Cooperation or Competition in the Caspian Sea Region: A Critical Review of the New Great Game Paradigm'. *Caucasus Survey*, 9(1): 1–20.

Bayramov, A. 2022. *Constructive Competition in the Caspian Sea Region.* Abingdon: Routledge.

Beaumont, Paul. 2024. *The Grammar of Status Competition: International Hierarchies and Domestic Politics.* Oxford: Oxford University Press.

Beaumont, Paul and Katharina Glaab. 2023. 'Everyday Migration Hierarchies: Negotiating the EU's Visa Regime'. *International Relations*, October. https://doi.org/10.1177/00471178231205408

Beaumont, Paul and Elana Wilson Rowe. 2022. 'Space, Nature and Hierarchy: The Ecosystemic Politics of the Caspian Sea'. *European Journal of International Relations*, December. https://doi.org/10.1177/13540661221142179

Becker, Bertha K. 1982. *Geopolitica da Amazonia: A Nova Fronteira de Recursos*. Rio de Janeiro: Zahar.

Bratman, Eve. 2019. *Governing the Rainforest: Sustainable Development Politics in the Brazilian Amazon*. Oxford: Oxford University Press.

Burke, Danita Catherine. 2019. *Diplomacy and the Arctic Council*. Montreal: McGill-Queen's University Press.

Cameron, Laura and Sinead Earley. 2015. 'The Ecosystem: Movements, Connections, Tensions and Translations'. *Geoforum* 65: 473–481.

Caraccioli, M. 2021. *Writing the New World: The Politics of Natural History in the Early Spanish Empire*. Gainesville, FL: University Press of Florida.

Church, Jon Marco. 2020. *Ecoregionalism: Analyzing Regional Environmental Agreements and Processes*. Abingdon: Routledge.

Corry, Olaf. 2024. 'Making the Climate Malleable? "Weak" and "Strong" Governance Objects and the Transformation of International Climate Politics'. *Global Studies Quarterly* 4(3): ksae062.

De Carvalho, Benjamin, Halvard Leira, Alejandro Colás, Maria Mälksoo and Mark Shirk, eds. 2022. *The Sea and International Relations*. Manchester: Manchester University Press.

Depledge, Duncan. 2018. *Britain and the Arctic*. Cham: Springer International Publishing.

Dodds, Klaus and Mark Nuttall. 2016. *The Scramble for the Poles*. Cambridge: Polity Press.

English, John. 2013. *Ice and Water: Politics, Peoples and the Arctic Council*. Toronto: Penguin Group (Canada).

Esguerra, Alejandro. 2024. 'Objects of Expertise: The Socio-material Politics of Expert Knowledge in Global Governance'. *Global Studies Quarterly* 4(3): ksae060.

Gad, Ulrik Pram and Jeppe Strandsbjerg, eds. 2019. *The Politics of Sustainability in the Arctic: Reconfiguring Identity, Space and Time*. Abingdon: Routledge.

Garcia, Beatriz. 2011. *The Amazon from an International Law Perspective*. Cambridge: Cambridge University Press.

Greer, Kirsten. 2015. 'Zoogeography and Imperial Defence: Tracing the Contours of the Nearctic Region in the Temperate North Atlantic, 1838–1880s'. *Geoforum* 65: 454–464.

GRID-Arendal. 2012. 'Protected Areas of the Caspian Sea'. Available at: https://www.grida.no/resources/5716

Gruby, Rebecca L. 2017. 'Macropolitics of Micronesia: Toward a Critical Theory of Regional Environmental Governance'. *Global Environmental Politics* 17(4): 9–27.

Heathershaw, J. and Megoran, N. 2011. 'Contesting Danger: A New Agenda for Policy and Scholarship on Central Asia'. *International Affairs*, 87: 589–612.

Lake, David A. 2007. 'Escape from the State of Nature: Authority and Hierarchy in World Politics'. *Insecurity Security*, 32(1): 47–79.

Lake, David A. 2009. *Hierarchy in International Relations*. Ithaca, NY: Cornell University Press.

Landau, Georges. 1981. 'El Tratado de Cooperac Ión Amazónica'. *Comercio Exterior* 31(12): 1386–1396.

Lehman, Jessica. 2020. 'Making an Anthropocene Ocean: Synoptic Geographies of the International Geophysical Year (1957–1958)'. *Annals of the American Association of Geographers* 110(3): 606–622.

Lewis, Martin W. and Kären Wigen. 1997. *The Myth of Continents: A Critique of Metageography*. Berkeley: University of California Press.

Little, Paul E. 2001. *Amazonia: Territorial Struggles on Perennial Frontiers*. Baltimore, MD: Johns Hopkins University Press.

MacDonald, Paul K. 2018. 'Embedded Authority: A Relational Network Approach to Hierarchy in World Politics'. *Review of International Studies* 44(1): 128–150.

Maglia, Cristiana and Elana Wilson Rowe. 2023. 'Ecosystems and Ordering: Exploring the Extent and Diversity of Ecosystem Governance'. *Global Studies Quarterly* 3(2): ksad028.

Maglia, Cristiana and Elana Wilson Rowe. 2024. 'Ecosystems and Ordering: a dataset'. *Data in Brief*: 111085.

Mattern, Janice Bially and Ayşe Zarakol. 2016. 'Hierarchies in World Politics'. *International Organization* 70(3): 623–654.

Mcconaughey, Meghan, Paul Musgrave and Daniel H. Nexon. 2018. 'Beyond Anarchy: Logics of Political Organization, Hierarchy and International Structure'. *International Theory* 10(2): 181–218.

Michelsen, Nicholas. nd. 'The International Relations of Tropical Storms in the Caribbean'. Available at: https://kclpure.kcl.ac.uk/portal/en/projects/the-international-relations-of-tropical-storms-in-the-caribbean

Mills, Eric. 2009. *The Fluid Envelope of Our Planet: How the Study of Ocean Currents Became a Science*. Toronto: University of Toronto Press.

Mills, Eric. 2011. *Biological Oceanography: An Early History. 1870–1960*. Toronto: University of Toronto Press.

Musgrave, Paul and Daniel H. Nexon. 2018. 'Defending Hierarchy from the Moon to the Indian Ocean: Symbolic Capital and Political Dominance in Early Modern China and the Cold War'. *International Organization* 72(3): 591–626.

Nunes, Paulo Henrique Faria. 2016. 'A Organização do Tratado de Cooperação Amazônica: Uma Análise Crítica das Razões por Trás da Sua Criação e Evolução'. *Revista de Direito Internacional* 13(2).

Paes, Lucas de Oliveira. 2022. 'The Amazon Rainforest and the Global–Regional Politics of Ecosystem Governance'. *International Affairs* 98(6): 2077–2097.

Paes, Lucas de Oliveira. 2023. 'Networked Territoriality: A Processual–Relational View on the Making (and Makings) of Regions in World Politics'. *Review of International Studies* 49(1): 53–82.

Paes, Lucas de Oliveira. 2025. 'Hierarchy', in Beate Jahn and Sebastian Schindler, eds, *Elgar Encyclopedia of International Relations*. Cheltenham: Edward Elgar, pp 159–160.

Paes, Lucas de Oliveira. Forthcoming. *Bargaining Stewardship: Amazonian Regionalism in Global Governance*.

Pereira, Joana C. and Eduardo Viola. 2020. 'Close to a Tipping Point? The Amazon and the Challenge of Sustainable Development under Growing Climate Pressures'. *Journal of Latin American Studies* 52(3): 467–494.

Pereira, Joana Castro and Eduardo Viola. 2019. 'Catastrophic Climate Risk and Brazilian Amazonian Politics and Policies: A New Research Agenda'. Global Environmental Politics 19(2): 93–103.

Pikialasorsuaq Commission. 2017. 'People of the Ice Bridge: The Future of the Pikialasorsuaq'. Available at: http://pikialasorsuaq.org/en/Resources/Reports

RAISG (Red Amazónica de Información Socioambiental Georreferenciada). 2022. Áreas Protegidas, Territorios Indígenas y Densidad de Carbono. Available at: https://www.raisg.org/es/publicacion/amazonia-2022-areas-protegidas-territorios-indigenas-y-densidad-de-carbono/

Raspotnik, Andreas. 2018. *The European Union and the Geopolitics of the Arctic*. Cheltenham: Edward Elgar.

Rekacewicz, Philippe, Riccardo Pravettoni and Nieves Lopez Izquierdo. 2019. 'Plastic Input into the Arctic Ocean Philippe Rekacewicz, Riccardo Pravettoni and Nieves Lopez Izquierdo'. *GRID-Arendal*. Available at: https://www.grida.no/resources/13350

Roszko, Edyta. 2022. 'Transoceania: Connecting the World beyond Eurasia', in Agnieszka Pasieka and Juraj Buzalka, eds, *Anthropology of Transformation*. Cambridge: Open Book Publishers, pp 221–242.

Rottem, Svein Vigeland. 2020. *The Arctic Council: Between Environmental Protection and Geopolitics*. New York: Palgrave Macmillan.

Scott, James C. 2020. *Seeing Like a State: How Certain Schemes to Improve the Human Condition Have Failed*. New Haven, CT: Yale University Press.

Shadian, Jessica. 2014. *The Politics of Arctic Sovereignty*. Abingdon: Routledge.

Sjoberg, Laura. 2017. 'Revealing International Hierarchy through Gender Lenses', in *Hierarchies in World Politics*. Cambridge: Cambridge University Press, pp 66–95.

Steinberg, Philip E., Jeremy Tasch, Hannes Gerhardt, Adam Keul and Elizabeth A Nyman. 2015. *Contesting the Arctic: Politics and Imaginaries in the Circumpolar North*. New York: I.B. Tauris.

Suzuki, Shogo. 2017. '"Delinquent Gangs" in the International System Hierarchy', in Ayşe Zarakol, ed., *Hierarchies in World Politics*. Cambridge: Cambridge University Press, pp 219–240.

Tigre, Maria Antonia. 2017. *Regional Cooperation in Amazonia: A Comparative Environmental Law Analysis*. Leiden: Brill.

Turnbull, David. 2005. *Masons, Tricksters and Cartographers: Comparative Studies in the Sociology of Scientific and Indigenous Knowledge*. London: Routledge.

Tussie, Diana. 2009. 'Latin America: Contrasting Motivations for Regional Projects'. *Review of International Studies* 35(S1): 169–188.

UNDP. 2008. *The Caspian Sea: Restoring Depleted Fisheries and Consolidation of a Permanent Regional Environmental Governance Framework 'CaspEco'*. UNDP Project Document. PIMS #4058 Governments of: Azerbaijan, Islamic Republic of Iran, Kazakhstan, Russian Federation & Turkmenistan. Available at: https://www.thegef.org/project/caspian-sea-restoring-depleted-fisheries-and-consolidation-permanent-regional-environmental

Viola, Eduardo and Matías Franchini. 2017. *Brazil and Climate Change: Beyond the Amazon*. Abingdon: Routledge.

Wilson Rowe, Elana. 2018. *Arctic Governance: Power in Cross-Border Relations*. Manchester: Manchester University Press.

Wilson Rowe, Elana. 2021. 'Ecosystemic Politics: Analyzing the Consequences of Speaking for Adjacent Nature on the Global Stage'. *Political Geography* 91: 102497.

Wilson Rowe, Elana. Forthcoming. *Earth in Pieces: Ecosystemic Politics and the Anthropocene*.

Wood-Donnelly, Corine. 2020. *Performing Arctic Sovereignty: Policy and Visual Narratives*. Abingdon: Routledge.

Yao, Joanne. 2019. '"Conquest from Barbarism": The Danube Commission, International Order and the Control of Nature as a Standard of Civilization'. *European Journal of International Relations* 25(2): 335–359.

Yao, Joanne. 2021. 'An International Hierarchy of Science: Conquest, Cooperation and the 1959 Antarctic Treaty System'. *European Journal of International Relations* 27(4): 995–1019.

Yao, Joanne. 2022. *The Ideal River: How Control of Nature Shaped the International Order*. Manchester: Manchester University Press.

Zarakol, Ayse, ed. 2017. *Hierarchies in World Politics*. Cambridge: Cambridge University Press.

5

To Unveil Nature's Secrets: International Cooperation in the International Geophysical Year

Joanne Yao

Introduction

The International Geophysical Year (IGY) of 1957–1958 and the international cooperative frameworks that emerged from activities conducted during the IGY, the 1959 Antarctica Treaty System and the 1967 Outer Space Treaty are often celebrated as examples of how science can help international society overcome divisive politics to achieve win-win scenarios for the benefit of all humanity. Indeed, scholarship on international environmental cooperation often looks to the conduct of science and appeals to scientific expertise to help states turn towards cooperation despite national self-interest and relative gains concerns. However, this chapter critically evaluates the IGY as a model for international cooperation and highlights how international cooperation rests on the unexamined assumption that cooperation between human societies to master nature will lead to a more peaceful and progressive world. To do so, I use a combination of archival diplomatic material and IGY promotional material in the form of film, art and print to focus on narratives used to justify the IGY to an international public.

This chapter interrogates the IGY as the embodiment of what I call 'epistemic completion', the scientific quest to complete our knowledge of the globe as an integrated whole. First, the chapter contends that international cooperation as a key aim of the liberal order relies on the promise that an international society that works together to master nature can secure win-win outcomes. Then, I introduce epistemic completion as a concept that enables international cooperation through the conviction that only by seeing and knowing from everywhere all at once can we truly master the globe for the benefit of all. The IGY as a quest for epistemic completion made precisely this argument that complete knowledge of the globe can only be acquired by building a globe-spanning network of scientists through international cooperation. I then show that the IGY is able to foster international cooperation through two interrelated narratives: first, the idea that science constituted an escape from divisive Cold War politics; and, second, that science represented a collective human adventure to master nature. However, as the final section of this chapter argues, international cooperation in the IGY also necessitated the material and ideational infrastructure of empire, and efforts to subsume global diversity into a single narrative of scientific progress legitimated and reinscribed colonial hierarchies under the banner of global scientific cooperation.

International cooperation and mastery of nature

Attention to Global Environmental Politics (GEP) within International Relations (IR) has traditionally been the domain of liberal institutionalism, with a focus on how to forge successful international cooperation and the political economy of the environment (for example, Vogler 2005; Young 2010; Haas 2015; Keohane 2015). While some notable exceptions foreground discourses in environmental agreement making (Litfin 1995; Epstein 2008), traditional GEP largely considers global environmental challenges to be a collective action problem that international actors must solve through collaboration. However, while mainstream GEP literature has been constructive in bringing environmental issues to the analytical forefront, a central tension rests at the heart of liberalism's relationship with nature that is underexplored by mainstream GEP literature. In particular, GEP tends to treat environmental challenges as a negative externality rather than coming to terms with the liberal international order's own complicity in creating those conditions. In this section, I explore how international cooperation hinges on the assumption that cooperative, peaceful and progressive relations between international actors can only be achieved through mastery of nature.

Since the beginning of IR as an academic area of study, disagreements between realists and liberals have centred on the potential for international progress. Realists from Hans Morgenthau to Kenneth Waltz and John Mearsheimer adopt pessimistic views of human nature and portray an international sphere governed by the unchanging principles of power and competition. For example, Morgenthau maintains in *Politics among Nations* that peace cannot be achieved by imposing limitations on national sovereignty because of the 'very nature of relations among nations' (1948: 419). Against this doubtful attitude towards progress, liberals draw from Immanuel Kant's proposition that there is a 'teleological theory of nature' where 'all natural capacities of a creature are destined to evolve completely to their natural end' (1784). This sense of progressive possibilities informs liberal thinkers of international politics who argue that norms, laws and institutions can moderate, if not overcome, our conflictual nature as individuals and anarchy in the international system (for example, Angell 1913; Keohane 1984; Ikenberry 2001). These narratives of progress often draw inspiration from global (but Western-led) histories of scientific and technological advancements that enable societies to become wealthier, freer and increasingly peaceful towards one another. However, this vision of progress rests on an ever-upwards trajectory of material abundance that relies on mastering the natural world and controlling nature for the benefit of human societies.

The liberal international order's reliance on dominating nature to fuel economic growth and societal progress has a long history. Before the 19th century, colonialism in the Americas consolidated European views that civilized societies worked to tame and improve the natural environment, while less advanced societies did not (Drayton 2000; Quijano 2000). For example, in his 1780 *Essays on the History of Mankind in Rude and Cultivated Ages*, Scottish philosopher James Dunbar defends colonial conquest by contending that the industry of civilized societies can transform nature and that 'by opening the soil, by clearing the forests, by cutting out passages from stagnant waters, the new hemisphere becomes auspicious, like the old, for the growth and population of mankind'. To achieve this progress, he concludes that we should 'learn then to wage war with the elements, not with our own kind' so that we can reclaim 'our patrimony from Chaos' (1780: 338). Here, Dunbar echoes John Locke's labour theory of property but adds an intersocietal dimension. He expresses the liberal argument that humanity should not engage in fratricidal violence, but should cooperate with one another to master their environment for the prosperity of all. Those who stood in the way sided with 'Chaos' against the civilized world. In the 19th century, taming nature became a civilizational standard that informed the globalization of European international society as states that could not control nature for global commerce were deemed less advanced (Yao 2019). These standards continued into the 20th and 21st centuries as legitimate statehood continues to be predicated on a state's ability to master nature and extract value from the resources within a defined and bounded territory (Scott 1998; Allan 2018; Yao 2022).

Early global governance also aimed to master the natural world. The first intergovernmental organizations and first formal institutions of global governance were established to manage transboundary rivers: the 1815 Rhine Commission and the 1856 Danube Commission. These international bodies forged cooperative frameworks to control the river and create a seamless highway for local and international trade, but they also made conquests from barbarism and facilitated progress along the metaphorical river of history (Yao 2022). These entwined goals of economic and moral progress continued to inform global governance in the 20th century as the victors of the Second World War established the United Nations and Bretton Woods systems as institutions of the liberal international order. However, this compact relied on mastery over nature – both through the science of thermodynamics which allowed the imperial powers to harness fossil fuels for capitalist accumulation and economic progress (Dagget 2019), and through mastery over atomic power that allowed the West to maintain military dominance (Biswas 2014). Into the 21st century, the promise of absolute gains and win-win economic outcomes for all continues to legitimate liberal institutions of global governance. However, climate change threatens this promise. As Amitav Ghosh sharply observes, it is not possible for everyone to have 'two cars, a washing machine, and a refrigerator – not because of technical or economic limitations but because humanity would asphyxiate in the process' (2016: 92–93). This quote highlights that the planetary challenges we face do not stem from external constraints that limit growth, but from our own unsustainable assumptions, or what Ghosh calls the 'great derangement', that have long underpinned global governance and the promise of liberal progress.

In IR, one specific post-Cold War debate further illustrates the unspoken importance of mastery over nature for international cooperation. Disagreements between structural realist and liberal institutionalists hinge on relative versus absolute gains concerns. In his often-cited attack on the 'false promise' of international institutionalism, John Mearsheimer outlines the difference – when a state considers absolute gains, it focuses 'on maximizing its own profit, and cares little about how much the other side gains or loses', but when it focuses on relative gains, it cares about both how much it gains and also how that gain compares to others' gain (1994: 12). Mearsheimer argues that institutionalists focus almost exclusively on absolute gains, but relative gains considerations make cooperation difficult under conditions of anarchy. In their response, liberals highlight that international institutions help resolve precisely relative gains concerns, what they term 'distributional issues', among not two but multiple states (Keohane and Martin 1995: 44). Institutions allow for coordination, information sharing and side-payments that can alleviate relative gains concerns. Much GEP scholarship focuses on these mechanisms and the role of scientific experts in forging them at international forums (Haas 1989; Young 2002; Daoudy 2009; Hale 2020). But in order to achieve increasing absolute gains and continue to maximize profits from international cooperation, international society must assume what Jason Moore frames as 'cheap nature' (2016) – that nature is a bountiful resource which, for the price of extraction and processing, could be transformed into endless economic growth. This assumption requires international society to tame nature and render it productive to ensure the 'win-win' outcomes that underpin the liberal international order.

The history of global governance demonstrates the liberal order's reliance on the implicit assumption that international progress is only achievable through the mastery of nature. Within IR, this assumption creates an opposition between a tragic world where states engage in fratricidal violence with one another, and a progressive international society under the united purpose of cooperation for the benefit of all. But how precisely does the quest to master nature translate into international cooperation? Controlling nature at the local level does produce material abundance through extraction and plantation economies, but by the late 19th and early 20th centuries, the world had become global, and a more totalizing framework was needed to produce ever more material gains. In the next section, I introduce the concept of epistemic completion and argue that in order to secure the material fruits of mastering nature, international cooperation in the service of the liberal order hinged on envisioning the globe as an integrated whole which necessitated collaboration between multiple actors to fully understand and control.

Epistemic completion and international scientific cooperation

In order to obtain maximum utility from the natural resources around us and ensure absolute gains for all, global powers must first strive to understand and master the Earth in its entirely. In this section, I argue that liberal international cooperation is inextricably linked to scientific discovery through what I call epistemic completion: the scientific quest to complete our knowledge of the globe as an integrated whole. This conceptualization is rooted in the rise of global sciences in the 19th century such as meteorology and oceanography that rests on a holistic vision of the Earth, and hence required the gathering of data from everywhere around the globe all at once. Doing so necessitated coordination between data collectors situated in different locations through international scientific cooperation.

The notion of epistemic competition is clearly articulated by a young John Ruskin, who would become one of the most influential Victorian critics and thinkers. In 1839, he wrote *Remarks on the Present State of Meteorological Science*, in which he argued for the formation of a meteorological society. In this short treatise, Ruskin outlined a vision of meteorology as a truly global science that differed from scientific projects that came before. Whereas before, 'a Galileo, or a Newton, by the unassisted workings of his solitary mind, may discover the secrets of the heavens', the meteorologist 'is impotent if alone'. Instead, understanding meteorology required more than just one individual genius in a laboratory or observatory; it required a globe-spanning network to collective data from everywhere all at once – a network that could 'think, observe, and act simultaneously'. Together, these observers would form 'a vast machine' and 'one mighty Mind – a ray of light entering into one vast Eye – a member of a multitudinous Power, contributing to knowledge and … solving the most deeply hidden problems of Nature' (Ruskin 1839: 56–59).

Here, Ruskin encapsulates the aims of epistemic completion. It is both a project to conquer the globe in its entirely by unveiling all of Nature's secrets before a globe-spanning network of scientists, as well as an act of creation to reconfigure the global commons as a space for an all-knowing and all-seeing science. This idea that data collection from around the globe is necessary to understand a systemic geophysical phenomenon extends beyond meteorology to other prominent 19th-century sciences including the study of oceanography, magnetism, geology and biological evolution. The quest to discover one last piece of the planetary puzzle spurred exploration to remote corners of the globe in the name of science, from the Galapagos Islands to the Arctic and Antarctica. The construction of Ruskin's 'vast Eye' to discover Nature's hidden secrets continued into the 20th and 21st centuries, particularly through the space age and our newfound ability to look back on the Earth from outside. This was most poetically captured by the 1968 *Earthrise* and 1972 *Blue Marble* images taken by astronauts from the Apollo 8 and 17 spacecrafts which portrayed the Earth as a beautiful, fragile, integrated whole. Into the 21st century, meteorologists have noted Ruskin's prescience as his vast machine is 'largely complete, built from parts – satellites, instantaneous telecommunications, and computers – that Ruskin could never have imagined' (Edwards 2004: 827). Humanity's ability to address the environmental challenges of the future still rests on the continual production and fine-tuning of Ruskin's one vast eye.

While the collective global quest for epistemic completion gathered pace in the 19th and 20th centuries, it has a longer history embedded in inter-imperial cooperation and competition over scientific and territorial conquest (Brockway 1979; Grove 1996; Allan 2018). Perhaps the most famous moment of inter-imperial scientific cooperation was the late 18th-century effort to determine the distance between the Earth and the sun, known as the Astronomical Unit, by observing the transit of Venus from multiple locations around the globe. The transit, where Venus passes in front of the sun, is a rare event that comes in pairs in quick succession, but the pairs are separated by more than a century. In 1714, renowned British astronomer Edmund Halley predicted the next transits will occur in 1761 and 1769 and argued that accurate measurements would be the single more important scientific advancement for humanity. While he would not live to observe the transits himself, Halley entreated fellow scientists to cooperate in this endeavour (Teets 2003; Wulf 2012). In 1760, French astronomer Joseph-Nicolas Delisle took up Halley's call and sent material about the transit to more than 200 scientists from Paris to St Petersburg to Constantinople. Despite the global

Seven Years' War and competition between European empires as to who was the most scientifically advanced, there was surprising cooperation. For example, French astronomer Alexandre Pingré carried with him documents from the British Admiralty stating that all British ships should allow Pingré to proceed without delays or interruptions (Wulf 2012: 15).

The second transit in 1769 attracted even more attention and funding from European empires that used their imperial networks to position astronomers in the most advantageous locations with the French in Pondicherry and Haiti, the British in the Hudson Bay and its American colonies, Sweden in Finland and Lapland (both part of the Swedish Empire at the time), and Denmark in the Arctic Circle in what is today Norway. The most famous such expedition was led by Captain James Cook and naturalist Joseph Banks whom the Royal Society sent to the South Pacific to measure the transit from Tahiti. In addition to its scientific duties, Cook's expedition also claimed Australia and New Zealand for the British crown. The observation locations also demonstrated inter-imperial cooperation – the French sent a team to Spanish territory in Baja California, while Russia's Catherine the Great welcomed German astronomers to the corners of her empire (Wulf 2012: 120). In these collaborations to observe the transit of Venus, we can see early indications of how the quest for epistemic completion through scientific discovery fosters international cooperation.

While international collaboration continued occasionally in pursuit of total scientific knowledge of global phenomena, for example, cooperation in terrestrial magnetism in the early 19th century (Cawood 1977), it was not until the 1957–1958 IGY that large-scale, globe-spanning scientific cooperation became self-consciously a project of epistemic completion and institutionalized as a project of global governance. In celebration of the momentous collaborative project, the September 1957 issue of the *UNESCO Courier*, a monthly magazine published in English, French, Spanish and Russian, dedicated the entire issue to the IGY. It begins by maintaining that while the planet's human cultures and environments are diverse, both are underpinned by 'an underlying unity' – there is an 'essential humanity' that connects men just as there are underlying scientific principles that animate nature. It is science's role to 'simplify and unify the diversity of nature' by revealing its essential unity. While individual scientists might be only able to make singular observations, 'the combined view of the human race, in the form of organized science, sees it whole' (1957). Here, the *Courier* expands Ruskin's notion of the all-seeing eye to a global programme of scientific enquiry designed to discover the hidden unity behind nature.

The creation of Ruskin's all-seeing eye as a basis of liberal international cooperation required two things. First, Ruskin's networks of scientists required inter-imperial cooperation to coordinate the gathering of data simultaneously from around the world all at once. In the next section,

I will explore how at the height of the Cold War, cooperation in the IGY was framed as an 'escape' from fractious geopolitics to enable humanity to master nature and understand the planet it its entirety. Second, the quest for epistemic completion depended on the ability to observe simultaneously from around the globe which relied on the material and ideational infrastructure of empire. In the final section, I will show how relying on this infrastructure legitimated and reinscribed imperial hierarchies in the mid-20th century under the banner of mastering nature for the benefit of all humanity.

The 1957–58 International Geophysical Year as an escape from politics

The IGY, which lasted 18 months from 1 July 1957 to 30 December 1958, was the most prominent example of how the quest for epistemic completion helped international actors overcome divisive geopolitics to achieve cooperation. The IGY was built on two previous international efforts – the 1882–1883 and 1932–1933 International Polar Years – but represented an expansion in scope, ambition and number of participants. A third polar year was first proposed in April 1950 at a dinner party hosted by US scientists for visiting British geophysicist Sydney Chapman and taken to the International Council of Scientific Unions (Belanger 2004: 483–484). Later, the World Meteorological Organization recommended the remit be expanded beyond the poles to the entire Earth and its surroundings (Collis and Dodds 2008: 556–557).

In the midst of the Cold War, the IGY was an impressive feat of international cooperation by states on both sides of the ideological divide. More than 60,000 scientists from 67 states joined the effort to collect and analyse scientific data and capture knowledge about the globe in its entirety (Korsmo 2007). The IGY's data-gathering efforts focused on sciences that required comprehensive data-gathering networks to form Ruskin's vast eye – from the study of auroras, cosmic ray, ionospheric physics and solar radiation to meteorology, oceanography, and seismology to geomagnetism, gravity and precision mapping. The IGY's ability to foster cooperation rested on two key narratives: first, the framing of science as an 'apolitical' aim for the good of all humanity that bypasses the messy everyday politics of the Cold War; and, second, the narration of the IGY's collective aim as an exciting human adventure and act of conquest that required human cooperation to gain mastery over nature.

Epistemic completion as escape from violent geopolitics

Films and press material surrounding the IGY romanticized the contrast between cooperative science and politics elsewhere. To launch the IGY at midnight on 1 July 1957, the BBC produced *The Restless Sphere*, where His Royal Highness Prince Philip, as a Fellow of the Royal Academy, introduced the IGY's globe-spanning activities. The broadcast begins with a stark contrast between war and peaceful cooperation with Prince Philip declaring: 'War – argument and controversy and news. Peaceful cooperation is a bore. I think that partially explains why so little has been heard about the International Geophysical Year until quite recently.' He goes on to explain why peaceful scientific cooperation is not so boring, and ends by coming back to science as peaceful cooperation: 'The IGY is the world studying itself. But it is also much more than that. It is a great experiment in world cooperation.' Of course, the IGY will collect scientific data to satisfy curiosity and usher in practical discoveries that will 'improve the material wellbeing of the peoples of this world', but more importantly, it might be the 'first tentative step to a friendlier and more tolerant feeling between the nations of the world' (1957). Here, Philip articulates the liberal proposition that cooperation in the service of science brings material progress as well as a more peaceful world. In a promotional *United Nations Educational, Scientific and Cultural Organization* (UNESCO) pamphlet outlining the IGY's scientific programme, German journalist Werner Buedeler echoes Prince Philip's juxtaposition of war and scientific cooperation, writing: 'Scientists of more than fifty nations ... are joining together for a united undertaking. Never before, except for war, have so many of them been mobilized for a common cause' (1957: Preface). By contrasting war and peaceful scientific cooperation, Werner highlights how the IGY's scientific programmes are key to diverting the world away from war towards a more peaceful common project.

The executive director of the US National Committee for the IGY Hugh Odishaw was more explicit in articulating the role science plays in overcoming fractious international politics. In the years before the IGY, Odishaw had led the creation of classroom and public relations material, and, in its aftermath, led the production of the film series *Planet Earth* that summarizes the IGY's outcomes (Korsmo 2004). In an article published in the *Journal of International Affairs* immediately after the IGY in 1959, Odishaw argues that the nongovernmental organization (NGO) overseeing the IGY was able to foster cooperation 'even in a period of sharp political unrest' because proceedings were 'largely devoid of political aspects: discussions therefore were generally objective, it was not difficult to arrive at agreements, and relations were amicable' (1959: 47 and 54). This method, Odishaw suggests, might be used 'for securing international cooperation in areas of human interest besides the geophysical sciences' (1959: 52). For Odishaw, focusing on the pursuit of objective scientific aims allowed international actors to escape contentious geopolitical disagreements and contribute towards the common quest for knowledge. Similarly, *New York Times* correspondent Walter Sullivan wrote in the *Bulletin of the Atomic Scientists* that the fact the IGY remained 'devoid of political entanglement' reflected how 'science, in treating our planet as indivisible, is far ahead of politics, which treats it as two worlds' (1959: 68). By uniting behind science, the IGY enabled nations to overcome divisive Cold War politics.

Participants at the 1955 Paris and Brussels Conferences to coordinate IGY activities also echoed the narrative that the IGY is an objective and scientific rather than political enterprise. For example, in a letter from Australian representative J.P. Shelton to P.G. Law in the Division of External Affairs from the 1955 Paris Conference to coordinate IGY activities in Antarctica, Shelton writes that 'political matters were barred completely and most effectively … the first resolution adopted stated that the purposes of the conference were exclusively scientific. All attempts to introduce political considerations were emphatically rejected by the Chairman' (NAA, A1838/450945). At the 1955 Brussels meeting of the IGY planning committee at which the Soviet Union presented its IGY plans, an Australian report notes that 'for their part, after satisfying themselves on the point of Russian co-operation, the Americans were happy to let the meeting develop as a fraternity of scientists. There was no breath of politics at any time' (NAA, A1838/450944). Part of the IGY's 'apolitical' characteristics can be attributed to its structure and the creation of an NGO, the Comité Spécial de l'Année Géophysique Internationale (CSAGI), to lead and coordinate activities (Belanger 2004). By focusing on supposedly objective scientific aims rather than Cold War politics, IGY activities allowed nations to escape the tension that hindered international cooperation elsewhere.

However, extensive scholarship has shown that IGY participation was deeply political, and while the IGY did foster unprecedented international cooperation, it also contributed to nationalistic preoccupations and parochial military objectives. The US and the Soviet Union competed to showcase who could conduct the most impressive science. Sputnik, launched during the IGY, was calculated to demonstrate the Soviet Union's superiority (Siddiqi 2000). The IGY's focus on international space cooperation helped make the US's space-based military and intelligence programmes more palatable and 'resistant to criticism' (Manzione 2000: 50). Smaller powers also saw nationalistic opportunities in the IGY. Richard Powell contends that the Canadian government's 1958 Polar Continental Shelf Project during the IGY was an attempt 'to mobilize a pan-Canadian nationalism in response to perceived American and Soviet incursions upon territorial sovereignty' (2008: 620). For newly independent states such as India, IGY science offered a way to assert themselves as modern and progressive international actors (Kochkar 2008).

In addition to the political and military aims that coexisted alongside peaceful cooperation during the IGY, the very notion that focusing on 'objective' science offered an escape from violent, militarized politics was in itself political in that it underpinned a liberal international vision of progress. In a speech given to the Royal Society by Secretary to the British IGY Committee Dr D.C. Martin, he concludes that the IGY's ability to foster cooperation between scientists of all nations presents a lesson and 'we can look forward to a time when the excitement of finding out, with the power for good such knowledge can bring, can take place on man's planet without the shadow of the fear of war' (Martin 1958: 28). Rather than apolitical, this narrative that the pursuit of scientific knowledge is a force for moral good and can unify all humanity informs a particular political conceptualization of international progress that is central to the liberal international order.

Epistemic completion as collective conquest

In addition to framing international scientific cooperation as an escape from politics, public information campaigns also portrayed the IGY as a fantastic adventure that would allow humanity to conquer nature and unmask its secrets for the benefit of humanity. Ahead of the first International Polar Year in the late 19th century, the Austrian explorer Karl Weyprecht, one of its key orchestrators, gave public lectures promoting the project by arguing that 'the key to many secrets of Nature' can be found in exploring the polar regions, but as long as the project was 'merely a sort of international steeple-chase, which is primarily to confer honor upon this flag or other ... these mysteries will remain unsolved'. Instead, solving Nature's mysteries required international cooperation to make consistent observations based on the same scientific method from around the globe (quoted in Baker 1982: 276). In his message launching the IGY in 1957, US President Dwight D. Eisenhower continued this narrative of adventure and conquest, calling the IGY 'one of the great scientific adventures of our time' that will yield knowledge to 'give us new understandings and new power over the forces of Nature' (Eisenhower 1957). This power, the conclusion of Werner Buedeler's UNESCO pamphlet notes, 'may provide humanity with the key to a complete understanding of the planet we live on. This is the promise of the International Geophysical Year' (1957: 68).

This sense of collective conquest saturates the artistic material associated with and developed to promote the IGY. For instance, at the conclusion of the BBC's *The Restless Sphere*, a narrator recites a quote from Christopher Marlowe's *Tamburlaine the Great* (1590) from which the programme derives its name:

> Nature, that framed us of four elements
> Warring within our breast for regiment
> Doth tech us all to have aspiring minds
> Our souls, whose faculties can comprehend

> The wondrous architecture of the world
> And measure every wandering planet's course
> Still climbing after knowledge infinite
> And always moving as the restless spheres
> Wills us to wear ourselves and never rest
> Until we reach the ripest fruit of all
> That perfect bliss and sole felicity
> The sweet fruition of an earthly crown.

The quote captures a sense of restlessness and insatiability as Tamburlaine's successes increase. Indeed, he suggests that he will not be satisfied until he achieves the 'earthly crown' and conquers the world in its entirely. The quote reflects the IGY spirit as well, not necessarily in territorial conquest, but in its restless efforts 'climbing after knowledge infinite' and the conquest of the Earth through knowledge. The IGY also inspired a series of national postage stamps, with many depicting satellites orbiting the Earth, scientific infrastructure, and wildlife such as penguins and polar bears. However, The American three-cent stamp features an iconic image from the Sistine Chapel where God is reaching out to touch the finger of Adam (Figure 5.1). Through this image, the stamp recalls the biblical moment when God raises man above nature: 'Be fruitful, and multiply, and replenish the Earth, and subdue it: and have dominion over the fish of the sea, and over the fowl of the air, and over every living thing that moveth upon the Earth' (Genesis 1:26–27). The depiction reinforces the idea that the IGY represents humanity's collective mastery of the planet.

Despite its focus on how IGY activities foster peace between human nations, promotional material also uses martial language to describe the confrontation between humanity and nature, imbuing the grand IGY adventure with a sense of very real conquest. For example, in the IGY edition of *The UNESCO Courier*, one article begins by calling IGY programmes a 'scientific attack on so complex a problem as the complete understanding of the Earth'. Later, a photo of Antarctica is captioned 'Secrets of Antarctica are now being discovered more rapidly as Nature's resistance is overcome with aid of modern machines' (1957). British geologist Sydney Chapman, president of the IGY organizing committee, wrote a publicity article in 1955 entitled 'Mass Attack on Earth's Mysteries' in which he compares the IGY's network of scientists to an army headed by 'scientific generals' (Good 2010: 186). Walter Sullivan, the *New York Times* journalist covering the IGY, titled his book on the enterprise *Assault on the Unknown*. In its pages, he often describes the IGY's efforts as an assault or attack on unknown geographies. For example, he describes the Special Committee on Oceanic Research's work as the 'development of a plan for a concentrated attack … on the least-known of great water areas, the Indian Ocean' (1961: 412). Of IGY

Figure 5.1: International Geophysical Year three-cent 1958 issue US stamp

Source: Wikimedia Commons, public domain

work in Antarctica, Sullivan writes that 'never in the history of exploration has there been in size, composition or scope of inquiry, an effort to compare with this international assault on a virtually unknown continent' (1961: 306). To win this battle, Sullivan maintains that international cooperation is vital and 'only together can the inhabitants of this planet learn its secrets and conquer its heartless inhospitality' (1961: 400). Here, he reinforces the liberal idea that international cooperation will allow a united humanity to fight and master nature.

The ability to master nature in its entirety holds several key advantages for international society to achieve win-win outcomes for the benefit of all. First, while planners highlighted scientific curiosity as the IGY's animating objective, the harnessing of scientific knowledge has the potential to yield economic gains. For example, Hugh Odishaw muses that perhaps humanity has been 'too pessimistic' about the limits of the planet's resources and that IGY activities may reveal that 'the earth may be richer than we have believed'. Mining and tilling the oceans will allow us to fully exploit this vast, untapped economic resource (1959: 5). Similarly, Buedeler's UNESCO pamphlet concludes with a confident speculation that IGY science may yet discover new resources such as 'uranium ores and many other mineral deposits, which, no doubt, will be found, in Antarctica' (1957: 67). Second, in addition to the potential economic bounty, the IGY also promises control over nature. In an article in *National Geographic*, the chairman of the US IGY committee Joseph Kaplan speculates how satellites sent into Earth's orbit to gather data is the first step in controlling the weather. He writes that 'concerning the dealings of man with nature: first, measure; then understand; then predict; and finally you may be able to control' (1957: 802). This sequence charts how scientific cooperation in pursuit of epistemic completion leads to mastery over the forces of nature.

While the IGY lasted only 18 months, its international political legacies continue to reinforce the idea that international scientific cooperation allows a united humanity to master nature. One of the institutional outcomes of the IGY was the establishment of the Antarctic Treaty System (ATS) at the 1959 Washington Conference. The US invited the 12 participants to the conference based on their scientific activities in Antarctica during the IGY.[1] The resulting agreement committed all parties to ensuring that the southern continent remained 'forever to be used for peaceful purposes' (ATS, Preamble) with a ban on the 'the carrying out of military maneuvers, as well as the testing of any type of weapons' (ATS, Article I). Diplomats negotiating the 1967 Outer Space Treaty consciously looked to the ATS as a model for forging an agreement that committed parties to frame 'outer space, including the moon and other celestial bodies' as 'the province of all mankind' (OST, Article 1). It commits parties to the use of the moon and other celestial bodies 'exclusively for peaceful purposes' and forbids the 'the establishment of military bases, installations and fortifications, the testing of any type of weapons and the conduct of military maneuvers' (OST, Article IV). These examples of international cooperation amid the political competition of the Cold War demonstrate the power of epistemic completion to foster international cooperation. Into the 21st century, the ATS continues to be upheld as a successful model for international cooperation. At its 50th anniversary, the then US Secretary of State Hillary Clinton lauded the ATS as 'a blueprint for the kind of international cooperation that will be needed more and more to address the challenges of the 21st century, and it is an example of smart power at its best' (Clinton 2009). Hence, while the IGY only lasted 18 months, its impact on global governance continued to reinforce the narrative that scientific cooperation can help humanity overcome divisive international politics and create a more peaceful planet.

Reinscribing global hierarchies

The IGY's globe-spanning data-gathering efforts aimed to see and understand the Earth as an integrated whole, which, promoters and globalists hoped, would engender a more peaceful and cooperative world. However, as I have already suggested, epistemic completion is not a neutral, apolitical project, but one in the service of a particular global and planetary order. Envisioning the planet through Ruskin's one vast eye created a disembodied and desocialized view from nowhere (Lazier 2011; Lekan 2014; Lehman 2020) or what Donna Haraway calls the 'god's eye' trick (1988). It is a totalizing view that allows a superior humanity to see all, know and control the Earth. In doing so, it produces what Denis Cosgrove calls the 'Apollonian gaze' that 'pulls diverse life on earth into a vision of unity … from a single perspective', which in reality constitutes an imperial master view underpinned by a 'universalizing teleology of Western Christianity' (2001: xi). In other words, epistemic completion produces a singular, commanding, scientific truth of the Earth that homogenizes the planet and its diversity in the service of a Western-led world order. The planet becomes just another object of investigation placed under the microscope of an all-powerful humanity that stands above and apart from it.

In this section, I contend that the IGY's construction of a united humanity that stands together to unveil nature's secrets conceals the unevenness and inequalities of colonial power. In what follows, I highlight how the material and ideational infrastructures of colonialism were central to the implementation and success of IGY projects, and yet, colonial hierarchies have not been explicitly recognized as a crucial aspect of the IGY. By obscuring the importance of empire in scientific cooperation and data collection, I suggest that prevailing narratives of the IGY as a unifying project to master nature helped to reinscribe the imperial hierarchies produced by the fading colonial empires of the mid-20th century.

The infrastructure of colonialism

In order to uncover Nature's secrets, IGY planner emphasized the importance of, first, gathering 'synoptic data' using the same methods everywhere and, second, ensuring data-collection efforts left as few blank spaces as possible. Combined, this would allow the IGY to accomplish what Joseph Kaplan, the chairman of the US National Committee, called the 'triangulation of the whole earth' (1954: 929). The ability to gather data using the same methods from everywhere all at once relied on the infrastructure of colonialism, and yet, the colonial hierarchies that allowed Western empires to effortlessly install data collection stations and train personnel to man remote outposts around there world was minimized in IGY discussions.

Planning documents ahead of the IGY emphasized both simultaneous collection using the same methods and complete coverage of the planet. For example, an Australian report stressed the importance of coordination in producing systematic data: 'a vital underlying idea of the holding of the International Geophysical Year is that observations made simultaneously in many parts of the world can yield incomparably more fruitful results than an equivalent quantity of observations scattered randomly' (NAA A1838/450933). A Chinese document emphasized the importance of a global network of stations: 'there must be a world-wide network of stations and observatories to acquire sufficient and reliable data. If any of these sections is left unattended to, the theory of the whole cannot be established' (NAA A1838/450926). Further, the South African weather bureau stressed the need to build more stations in the southern hemisphere because 'in the years to come, when the meteorological data collected are studied and analyzed it will be a matter of permanent regret if certain huge areas are found to be entirely "blank"' (NAA A1838/450933). What was left unsaid in these discussions was that both synchronicity and completeness would rely on the infrastructure of empire.

To achieve this totalizing vision, IGY programmes needed data-gathering stations situated around the world, and major participants relied on colonial infrastructure to establish the extensive network of stations which would otherwise be incomplete and full of blanks. During planning meetings ahead of the IGY, participating countries reported the location of their data stations which would send data to a central coordinating location. These reports reveal the deep vein of imperial infrastructure necessary for the IGY's quest for epistemic completion. For example, at the 1955 Brussels planning conference, the British report listed all its proposed stations including the collection of areological measurements from Malta, Adan, Gibraltar, Cyprus, Bahrein, Habbaniyah in Iraq, and the Falkland Islands; the collection of geomagnetism data from Singapore, Ibadan in Nigeria, British Antarctic claims and the Falkland Islands; the collection of solar activities data from the Royal Observatory in the Cape of Good Hope; and the collection of oceanography data from Freetown in the West African Coast (today the capital of Sierra Leone), Takoradi in the Gold Coast (in today's Ghana) and a series of British naval vessels stationed globally. These locations were reported in addition to data locations in Britain itself from Greenwich Observatory to Eskdalemuir in Scotland (FO 371/113962 A1522/51). While this may be read as an objective list of the best data-collection locations to fulfil the IGY's mission, it also demonstrates how deeply colonial power and hierarchies run through the IGY and were central to its data-collection efforts.

The British were not alone in leveraging their colonial infrastructure in the service of global science. In a global list of IGY data up to June 1958 in the New Zealand archives, colonial infrastructure runs through the document like a bright thread and knits together IGY's expansive global vision. For example, the four stations listed under Belgium – Uvira, Astrida, Lwiro and Ramangabo – are all located in the Belgian Congo. The last two stations on Australia's list – Rabaul and Madang – are in Papua New Guinea, an Australian colony until 1975. The French list includes Alger and Tamanrasset in Algeria, Ksara in Lebanon, M'Bour in Senegal, Lome in Togo, and Noumea in New Caledonia in the South Pacific. The Portuguese list include Lorenzo Marques, which was the name of Maputo, the capital of Mozambique until 1976. The US list includes Guam, Truk in the Federated States of Micronesia, and Koror in Palau. The Soviet Union's list includes Semipalatinsk in Kazakhstan, Askhabad in Turkmenistan, Khorog in Tajikistan, Kishinev in Moldova, Goris in Armenia, and Simferopol on the Crimean Peninsula (in what is still internationally recognized as Ukraine) (NZ Archives, R1127950). Empire was so central to the IGY's efforts that even one of the Netherland's commemorative stamps for the IGY features an outline of the Netherland Antilles, one of its colonies in the Caribbean. Hence, despite the IGY's narrative of its data-collection efforts as apolitical in the service of a united humanity, the new era of simultaneous and coordinated science depended to a great degree on the physical infrastructure of empire.

Narrative of civilization progress

In addition to the physical infrastructure of colonialism, the IGY also relied on the narrative of civilizational progress to legitimate its activities. In this narrative, scientific exploration is a progressive force that helps mankind advance from a past cloaked in barbarism into a civilized modernity where societies cooperated to uncover Nature's secrets.

Historically, scientific discovery has long been implicated in colonial conquest. During early colonialism, Europeans used differences in tools and technologies as a means of comparing themselves to the societies they encountered, and in the 18th and 19th centuries, science and technology became a marker of civilizational advancement that both 'justified Europe's global hegemony and vitally influenced the ways in which European power was exercised' (Adas 1989: 3–4). Colonial conquest opened up new geographies for scientific discovery just as scientific discovery also helped legitimate colonial expansion as valuable for human knowledge (Schiebinger and Swan 2005; Yao 2021). For example, the prominence of renowned botanist Joseph Banks on Captain James Cook's expedition to the South Pacific burnished the voyage's scientific credentials and helped legitimate the founding of Australia and New Zealand as British colonies.

In addition, colonial conquest helped constitute a particular notion of science that is universal, rational and a measure of civilizational superiority. Indeed, as Jodi Byrd argues, the enterprise to measure the transit of Venus in the late 18th century 'moved European conquest towards notions of imperialist planetarity that provided the basis for Enlightenment liberalism. The imperial planetarity that sparked scientific rationalism and inspired humanist articulations of freedom, sovereignty, and equality touched four continents and a sea of islands in order to cohere itself' (Byrd 2011: xx–xxi). The IGY, as the intellectual successor of these instances of early scientific cooperation, also took on its colonial legacies, particularly in elevating science as a marker of civilizational advancement that separates modern man from backwards societies. IGY narratives then obscured these hierarchies by placing all societies on the same developmental continuum with scientific modernity at its most advanced position.

For example, a film summarizing Australia's contributions to the IGY created by the Commonwealth Film Unit in 1959 begins its narrative with an introduction to the IGY and an effort to justify its ambitions:

we live in a physical world – a world of infinite variety. From this world, man has wrested a living – food, clothing, shelter – the means to a more comfortable life … for many years, man took the world for granted, peopling it with gods and spirit to explain its phenomena. But, with the development of civilization came time to satisfy man's restless curiosity. Because of its continual influence on our way of living, man has always been curious about what makes the world tick. (NFSA Films)

Here, the film pivots to the IGY and all the scientific projects that constitute a 'vast and coordinated scientific exploration of our environment'. In the video, the visual representation is revealing. When the narrator speaks of gods and spirits, the video depicts Indigenous Australians in canoes, the only time these populations are featured in the film. When the narrator pivots to mankind's restless curiosity and the IGY, the video depicts industry and white (mostly male) Australian scientists at work. Here, the film clearly portrays Western science as a progressive force that guides societies from a past where gods dominate the human imagination to a modernity where rationality and curiosity drive our engagement with the natural world. The video's depiction of Indigenous Australians subtly speaks to the enduring colonial and racialized assumption that underpin the IGY's narrative of scientific progress.

Further, IGY promotional material and press coverage also reinforce a narrative of historical progress that places science at the forefront, leading humanity into a more enlightened and prosperous future. Casting the IGY as a continuation of voyages of discovery from Christopher Columbus and scientific greats of the European canon perpetuates a sense that the IGY's quest for epistemic completion is being created by the West for the West. For instance, Hugh Odishaw compares the energy and excitement of the IGY as a scientific adventure in search of new frontiers to Columbus' discovery of America as 'adventures on strange waters and unknown lands' (1959: 56). In his book *Assault on the Unknown*, Walter Sullivan traces the intellectual origins of the IGY not only back to the two previous international polar years, but also back through Copernicus, Francis Bacon and Pythagoras in envisioning the Earth as a sphere and an integrated whole. These stories not only draw a historical connection between key moments of Western scientific discovery and the IGY, underscoring the progressive character of epistemic completion, but in doing so, also sideline other ways of seeing and knowing the globe. As Sheila Jasanoff observes, 'science, with its handmaiden technology, entered into the manufacture of globalization, making interconnectedness seem inevitable, while rendering invisible those forms of life that did not cohere with science's mode of self-understanding or were inimical to technoscientific notions of progress and development'

(2021: 840). At the moment of international decolonization in the mid-20th century, projects such as the IGY helped recast colonial hierarchies into the language of scientific progress.

Conclusion

The 1957–1958 IGY represented an unparalleled moment of international scientific cooperation despite the geopolitical tensions of the Cold War. Its successes translated into key international cooperative frameworks of the mid-20th century, including the 1959 ATS and the 1967 Outer Space Treaty, that demonstrated the potential for scientific collaboration to help international actors overcome divisive politics. In this chapter, I have argued that the IGY embodied the aim of what I term 'epistemic completion' – the scientific quest to complete our knowledge of the globe as an integrated whole. It is both a quest to know the Earth in its entirety by unmasking its secrets and an act of constitution to envision the global commons as the domain of an 'all-seeing' science. This vision of the globe continues to inform not just our understanding of planetary systems but also our understanding of our place in the cosmos and the potential for science to foster political cooperation as we face unprecedented environmental challenges.

This chapter has demonstrated that by focusing on the need to observe the globe from everywhere all at once, the IGY as a quest for epistemic completion was able to foster international scientific cooperation. It did so through two interrelated narratives – by framing the IGY's scientific programme as an escape from everyday Cold War politics and by narrating it as a collective human adventure to master nature. Hence, the IGY was not just a large-scale and coordinated data-gathering venture, but also a project to know and control the planet for the benefit of all and to advance human progress. A detailed examination of IGY narratives reveals how the collective quest to master nature underpins the current liberal international order and its promise of win-win cooperative outcomes and material abundance. However, the final section of this chapter also highlighted how the IGY as the quest for epistemic completion relies on the material and ideational infrastructure of empire. By subsuming colonial hierarchies in its narrative of scientific progress, epistemic completion legitimates and reinscribes these hierarchies into the 'progress and prosperity for all humanity' narrative at the heart of the 20th-century liberal order. This unrecognized incorporation of colonial and racialized hierarchies has implications for global efforts to confront the enduring legacies of colonialism, and, in particular, projects for environmental and planetary justice.

Note

[1] The 12 participants were Argentina, Australia, Belgium, Chile, France, Japan, Norway, New Zealand, the Soviet Union, South Africa, the UK and the US.

References

Adas, M. (1989) *Machines as the Measure of Men: Science, Technology and Ideologies of Western Dominance*. Ithaca, NY: Cornell University Press.

Allan, B. (2018) *Scientific Cosmology and International Orders*. Cambridge: Cambridge University Press.

Angell, N. (1913) *The Great Illusion: A Study of the Relation of Military Power to National Advantage*, 3rd edn. New York: GP Putnam's Sons.

'Australia in the International Geophysical Year' (1959) Produced by the Commonwealth Film Unit. Available at: https://www.youtube.com/watch?v=f1HTmXmTwX0&t=732s

Baker, F.W.G. (1982) 'The First International Polar Year, 1882–83', *Polar Record*, 21(132), 275–285.

Belanger, D. (2004) 'The International Geophysical Year in Antarctica: Uncommon Collaborations, Unprecedented Results', *Journal of Government Information*, 30(4), 482–489.

Biswas, S. (2014) *Nuclear Desire: Power and the Postcolonial Nuclear Order*. Minneapolis: University of Minnesota Press.

Brockway, L. (1979) *Science and Colonial Expansion: The Role of the British Royal Botanic Gardens*. New Haven: Yale University Press.

Buedeler, W. (1957) *The International Geophysical Year. UNESCO and Its Programmes*. Available at: https://unesdoc.unesco.org/ark:/48223/pf0000128401

Byrd, J. (2011) *The Transit of Empire: Indigenous Critiques of Colonialism*. Minneapolis: University of Minnesota Press.

Cawood, J. (1977) 'Terrestrial Magnetism and the Development of International Collaboration in the Early Nineteenth Century', *Annals of Science*, 34(6), 551–587.

Clinton, H. (2009) 'Remarks at the Joint Session of the Antarctic Treaty Consultative Meeting and the Arctic Council, 50th Anniversary of the Antarctic Treaty'. Available at: https://2009-2017.state.gov/secretary/20092013clinton/rm/2009a/04/121314.htm

Collis, C. and Dodds, K. (2008) 'Assault on the Unknown: The Historical and Political Geographies of the International Geophysical Year (1957–8)', *Journal of Historical Geography*, 34(4), 555–573.

Cosgrove, D. (2001) *Apollo's Eye: A Cartographic Genealogy of the Earth in the Western Imagination*. Baltimore: Johns Hopkins University Press.

Dagget, C. (2019) *The Birth of Energy: Fossil Fuels, Thermodynamics, and the Politics of Work*. Durham, NC: Duke University Press.

Daoudy, M. (2009) 'Asymmetric Power: Negotiating water in the Euphrates and Tigris', *International Negotiations*, 14(2), 361–391.

Department of Scientific and Industrial Research, R1127950 (1958) Antarctica – IGY [International Geophysical Year] Vol 2, Box 21, Archives New Zealand Te rua mahara o te kāwanatanga, Christchurch.

Drayton, R. (2000) *Nature's Government: Science, Imperial Britain, and the 'Improvement' of the World.* New Haven: Yale University Press.

Dunbar, J. (1780) *Essays on the History of Mankind in Rude and Cultivated Ages.* London: W. Strahan.

Edwards, P. (2004) '"A Vast Machine": Standards as Social Technology', *Science*, 304(5672), 827–828.

Eisenhower, D. (1957) 'Remarks by the President in Connection with the Opening of the International Geophysical Year', 30 June. Available at: https://www.eisenhowerlibrary.gov/research/online-documents/international-geophysical-year-igy

Epstein, C. (2008) *The Power of Words in International Relations: Birth of an Anti-Whaling Discourse.* Cambridge, MA: MIT Press.

FO 371/113962 (1955) 'International Geophysical Year (1957 to 1958)'. Foreign Office Papers, British National Archives, Kew, UK.

Ghosh, A. (2016) *The Great Derangement: Climate Change and the Unthinkable.* Chicago: University of Chicago Press.

Good, G. (2010) 'Sydney Chapman: Dynamo behind the International Geophysical Year', in R.D. Launius et al (eds) *Globalizing Polar Science: Reconsidering the International Polar and Geophysical Years.* New York: Palgrave Macmillan, pp 177–204.

Grove, R. (1996) *Green Imperialism: Colonial Expansion, Tropical Island Edens and the Origins of Environmentalism, 1600–1860.* Cambridge: Cambridge University Press.

Haas, P. (1989) 'Do Regimes Matter? Epistemic Communities and Mediterranean Pollution Control', *International Organization*, 43(3), 377–403.

Haas, P. (2015) *Epistemic Communities, Constructivism, and International Environmental Politics.* New York: Routledge.

Hale, T. (2020) 'Catalytic Cooperation', *Global Environmental Politics*, 20(4), 73–98.

Haraway, D. (1988) 'Situated Knowledge: The Science Question in Feminism and the Privilege of Partial Perspective', *Feminist Studies*, 14(3), 575–599.

Ikenberry, G.J. (2001) *After Victory: Institutions, Strategic Restraint, and the Rebuilding of Order after Major Wars.* Princeton: Princeton University Press.

Jasanoff, S. (2021) 'Humility in the Anthropocene', *Globalizations*, 18(6), 839–853.

Kant, I. (1784) *Idea for a Universal History from a Cosmopolitan Point of View.* Translated by L.W. Beck. Available at: https://www.marxists.org/reference/subject/ethics/kant/universal-history.htm

Kaplan, J. (1954) 'The Scientific Program of the International Geophysical Year'. *Proceedings of the National Academy of Sciences of the United States of America*, 40(10): 926–931.

Kaplan, J. (1957) 'How Man-Made Satellites Can Affect Our Lives', *National Geographic Magazine*, December, 791–810.

Keohane, R. (1984) *After Hegemony: Cooperation and Discord in the World Political Economy*. Princeton: Princeton University Press.

Keohane, R. (2015) 'The Global Politics of Climate Change: Challenge for Political Science', *PS: Political Science & Politics*, 48(1), 19–26.

Keohane, R. and Martin, L. (1995) 'The Promise of Institutionalist Theory', *International Security*, 20(1), 39–51.

Korsmo, F. (2004) 'Shaping up Planet Earth: The International Geophysical Year (1957–1958) and Communicating Science through Print and Film Media', *Science Communication*, 26(2), 162–187.

Korsmo, F. (2007) 'The Birth of the International Geophysical Year', *The Leading Edge*. Available at: https://web.archive.org/web/20170808232957id_/http://people.uncw.edu/emslies/documents/Korsmo2007birthofIGY.pdf

Kochkar, R. (2008) 'Science as a Symbol of New Nationhood: India and the International Geophysical Year 1957–58', *Current Science*, 94(6), 813–816.

Lazier, B. (2011) 'Earthrise; or, the Globalization of the World Picture', *American Historical Review*, 116(3), 602–630.

Lehman, J. (2020) 'Making an Anthropocene Ocean: Synoptic Geographies of the International Geophysical Year (1957–1958)', *Annals of the American Association of Geographers*, 110(3), 606–622.

Lekan, T. (2014) 'Fractal Earth: Visualizing the Global Environment in the Anthropocene', *Environmental Humanities*, 5(1), 171–201.

Litfin, K. (1995) *Ozone Discourses: Science and Politics in Global Environmental Cooperation*. New York: Columbia University Press.

Manzione, J. (2000) 'Legacy of Scientific Internationalism', *Diplomatic History*, 24(1), 21–24.

Martin, D.C. (1958) 'The International Geophysical Year', *Geographical Journal*, 124(1), 18–29.

Mearsheimer, J. (1994) 'The False Promise of International Institutions', *International Security*, 19(3), 5–49.

Moore, J. (2016) 'The Rise of Cheap Nature', in J. Moore (ed.) *Anthropocene or Capitalocene? Nature, History, and the Crisis of Capitalism*. Oakland: PM Press, 78–115.

Morgenthau, H. (1948) *Politics among Nations: The Struggle for Power and Peace*, 4th edn. New York: Alfred A. Knopf.

National Archives of Australia (NAA) A1838/450926 (1955) 'Antarctica – International Geophysical Year – General'.

National Archives of Australia (NAA) A1838/450944 (1955) 'Antarctica – International Geophysical Year – Brussels Conference 1955'.

National Archives of Australia (NAA) A1838/450945 (1955) 'Antarctica – International Geophysical Year – Paris Conference 1955'.

National Archives of Australia (NAA) A1838/450933 (1956) 'Antarctica IGY [International Geophysical Year] Paris Conference 1956'.

Odishaw, H. (1959) 'The International Geophysical Year and World Politics', *Journal of International Affairs*, 13(1), 47–56.

Powell, R.C. (2008) 'Science, Sovereignty and Nation: Canada and the Legacy of the International Geophysical Year, 1957–1958', *Journal of Historical Geography*, 34(4), 618–638.

Quijano, A. (2000) 'Coloniality of Power and Eurocentrism in Latin America', *International Sociology*, 15(2), 215–232.

The Restless Sphere (1957) Produced by BBC. 30 June. Available at: https://www.youtube.com/watch?v=EqzS7HToH2E

Ruskin, J (1839) 'Meteorology', in *On the Old Road, Volume 2, A Collection of Miscellaneous Essays and Articles on Art and Literature*. New York & Chicago: National Library Association. Available at: https://www.gutenberg.org/files/21263/21263-h/21263-h.htm#Page_153

Schiebinger, L. and Swan, C. (2005) *Colonial Botany: Science, Commerce, and Politics in the Early Modern World*. Philadelphia: University of Pennsylvania Press.

Scott J.C. (1998) *Seeing Like a State: How Certain Schemes to Improve the Human Condition Have Failed*. New Haven: Yale University Press.

Secretariat of the Antarctic Treaty (1959) 'The Antarctic Treaty (ATS)'. Available at: https://www.ats.aq/e/antarctictreaty.html

Siddiqi, A. (2000) 'Korolev, Sputnik, and the International Geophysical Year', in R. Launius et al (eds) *Reconsidering Sputnik: Forty Years since the Soviet Satellite*. Reading, MA: Harwood Academic, pp 43–72.

Sullivan, W. (1961) *Assault on the Unknown: The International Geophysical Year*. New York: McGraw-Hill.

Teets, D. (2003) 'Transit of Venus and the Astronomical Unit', *Mathematics Magazine*, 76(5), 335–348.

United Nations Educational, Scientific and Cultural Organization (UNESCO) (1957) 'The UNESCO Courier: International Geophysical Year', September. Available at: https://unesdoc.unesco.org/ark:/48223/pf0000068057

United Nations General Assembly (1966) 'Treaty on Principles Governing the Activities of States in the Exploration and Use of Outer Space, including the Moon and Other Celestial Bodies'. Available at: https://www.unoosa.org/oosa/en/ourwork/spacelaw/treaties/outerspacetreaty.html

Vogler, J. (2005) 'The European Contribution to Global Environmental Governance', *International Affairs*, 81(4), 835–850.

Wulf, A. (2012) *Chasing Venus: The Race to Measure the Heavens*. London: Windmill Books.

Yao, J. (2019) '"Conquest from Barbarism": The Danube Commission, International Order and the Control of Nature as a Standard of Civilization', *European Journal of International Relations*, 25(2), 335–359.

Yao, J. (2021) 'An International Hierarchy of Science: Conquest, Cooperation, and the 1959 Antarctic Treaty System', *European Journal of International Relations*, 27(4), 995–1019.

Yao, J. (2022) *The Ideal River: How Control of Nature Shaped the International Order*. Manchester: Manchester University Press.

Young, O. (2010) *Institutional Dynamics: Emergent Patterns in International Environmental Governance*. Cambridge, MA: MIT Press.

Young, O. (2002) *The Institutional Dimensions of Environmental Change: Fit, Interplay, and Scale*. Cambridge, MA: MIT Press.

6

Outer Space and Sovereignty in Post-Planetary Politics

Katharina Glaab

Introduction

Outer space usually does not feature in considerations of environmental politics; planetary politics stops with the invisible atmospheric border. This is curious as developments in outer space have implications for Earth and vice versa. In light of a new space race that expands economic activities to outer space to extract resources, develop industries or undertake space tourism while at the same time building infrastructures on Earth, questions arise as to how these space activities will impact sustainable development on Earth and in outer space. With the rising number of satellites in low Earth orbit and growing awareness of the issue of space debris and resource use, the sustainability of space activities has become an important concern for public and private actors, leading to calls for 'space environmentalism' and new regulatory initiatives that take seriously the notion that the planetary realm forms an additional ecosystem (Morin and Richard 2021; Lawrence et al 2022; Yap and Truffer 2022). However, while states manage environmental challenges on Earth within their own national jurisdiction or through multilateral cooperation, this model of environmental governance cannot simply be transferred to outer space and its related sustainability challenges. The terrestrial sovereignty arrangements that underlie today's global governance system do not apply, since no single state can establish authority about outer space. So what does it mean when efforts to govern nature extend beyond planetary boundaries and include outer space? This chapter will explore this question by focusing on the implications for our understanding of sovereignty in International Relations (IR).

IR scholarship has mainly engaged with outer space politics with a geopolitics or global commons framing which set the stage for discussions of the current state of space politics. They depict two dominant yet differing 'strategic ontologies' of outer space which define the 'subjects, objects, and relationships that constitute the international system' (Lerner and O'Loughlin 2023) and their priorities, as well as their understanding of sovereignty. A geopolitical ontology of outer space conceptualizes outer space as an extension of earthly geopolitics and power rivalries on Earth, while a global commons ontology of outer space foregrounds cooperation among states to manage the resources of outer space sustainably. These ontologies matter as they prescribe policy discourses on the future development of outer space and its governance. Yet, the empirically observable extension and transformation of space activities, which sees increasing private and state-driven activities to further space exploration and a space economy, can hardly be grasped in the traditional language of IR. Indeed, 'the language of the global seems somewhat ill equipped to come to terms with the ways in which the outer-Earth and other extraterrestrial spaces are already part of our everyday lives' (MacDonald 2007: 599). Tying in with recent calls that it is necessary to 'retool IR theory to confront an extraterrestrial political future' (van Wingerden and Vigneswaran 2024: 600), this chapter explores the role of nature in different ideal-typical ontologies of outer space. Taking its cue from recent interventions by scholars suggesting a 'planetary politics' that highlights the entanglement of humans, nature and species, and the shared responsibility for our planet, in this chapter I will argue that we are now entering the age of post-planetary politics. A post-planetary ontology of outer space brings forward relational thinking where the boundaries between the planet and outer space become increasingly blurred. Thinking relationally beyond Earth requires and allows for conceptual rethinking of sovereignty as a key concept in IR, and allows us to further push and extend our understanding of the relationship of nature and humans, as well as Earth and outer space. In contrast to the other chapters in this book, this chapter is not about transformations taking place on Earth, but the political relevance of transformations that are happening along the boundary and intersection of planet Earth and outer space, and considers the implications of these processes for future environmental governance and our understanding of sovereignty.

The chapter will first critically discuss the geopolitical ontology of outer space and the reflex of traditional IR scholarship to explain conflicts as a matter of geopolitics based on a simple understanding of sovereignty and power politics. In this reading, nature often only features as a resource that can be extracted and acquired, or territory that can be gained. In a second step, it will show how a global commons ontology of outer space opens the door for liberal perspectives on collective action and how sovereignty claims are limited through current outer space treaties. Although this perspective accounts for environmental issues in outer space, I will go one step further and argue for a post-planetary approach to propose a governance of outer space which would challenge traditional conceptualizations of sovereignty. This shows how a relational ontology helps not only to conceptualize Earth and outer space as co-constitutive, but also disrupts binaries of not only nature/human but also Earth/space and human/nonhuman. It shows that governance that aims at regulating environmental issues related to space activities such as the use of its environment and resources, cannot be seen as separated from broader ethical questions of future multi-planetary relations.

Outer space as geopolitics

In IR, outer space has at best been of marginal concern to theorists. Ever since the advent of the first national space programmes, outer space has been a small part of broader realist-framed discussions around international security concerns and great power politics. The development of space technologies during the Cold War period and the race to the Moon was an important part of the political-ideological conflict between the US and the Soviet Union. Which state would have the first man or woman in outer space, or which state would get humans to the Moon first was a decisive driver of the respective countries space programmes. However, the resulting space race, which was supported by massive financial resources and political efforts, was not only part of a quest for international status and symbolic dominance (Musgrave and Nexon 2018); space development also manifested an understanding of outer space as an area of geopolitical contest which conceptualized outer space as a strategically important region. Proponents of so-called 'astropolitics' – 'the study of the relationship between outer space terrain and technology and the development of political and military policy and strategy' (Dolman 2002: 15) – argue that the control of outer space and its resources would manifest territorial dominance on Earth. Great power conflicts and competition on Earth (as witnessed during the Cold War) would only be extended to outer space (Deudney 2020). Accordingly, those states that can control low-Earth orbit (LEO) are able to establish 'space power' or, as Everett Dolman, one of the key proponents of astropolitics put it: 'Who controls low-earth orbit controls near-Earth space. Who controls near-Earth space dominates Terra. Who dominates Terra determines the destiny of humankind' (Dolman 2002: 8).

This kind of thinking aligns with traditional understandings of sovereignty in IR. Clearly set around the idea of territorial boundaries, sovereignty in this rendering of outer space reflects the notion of power through territorial control when discussing control of geopolitical advantageous points in Earth's orbit. However, it also points to the practice of expanding the boundaries of sovereign states as a means of extending influence and securing their control.

After all, claims to authority over territory have a history in the colonial pasts of many countries and which have legitimized the establishment of unequal power systems. Duvall and Havercroft (2008) have pointed to the potential consequences of this realist strategy to claim authority and extend national sovereignty beyond traditional state boundaries into outer space. Accordingly, particularly the development of space weapon technologies depicts a new structure of imperial power. They argue that the control of orbital space and the ability of the United States to establish a missile defence system would manifest the sovereignty of the US and similarly diminish the sovereignty of other states through their 'ability to maintain territorial integrity by deterring enemies from attacking' (Duvall and Havercroft 2008: 764). This would not only extend US sovereignty into outer space but would also establish a capitalist sovereignty by restricting access to the territory itself and the new space economy, thereby establishing a new form of empire (Duvall and Havercroft 2008: 765). Building on Dolman's argument, they claim that this space-based empire would also lead to a reversal of territoriality as the necessity to control territory moves from Earth to outer space (Duvall and Havercroft 2008: 772). Moreover, it would challenge sovereignty as a constitutive principle of the structure of the international system, as subjects of such an empire cannot be sovereign (Havercroft and Duvall 2009: 47).

Critical geographers have long criticized this realist ontology of outer space and have warned that 'terrestrial geopolitics are increasingly being determined by extraterrestrial strategic considerations' (MacDonald 2007: 594), and that the geopolitical knowledge of outer space is similarly constitutive of outer space strategies and policies. This is problematic as geopolitical strategies mobilize representations of outer space that are dominated by imperialist and realist discourses, while neglecting other ways of understanding the development of outer space. The realist ontology presumes, for instance, an understanding of outer space as a *terra nullius* or a new frontier where geopolitical control over an empty territory can be established. Redfield (2002) showed how space exploration relied on the same narratives as earlier imperial endeavours. In this way, they not only repeat colonial narratives and geopolitical thinking which treat outer space as territory to be seized and exploited (Havercroft and Duvall 2009: 46), but also enable authority claims without being challenged. From an Anthropocentric point of view, one could argue that a key difference to former colonial projects is that today's space exploration covers territory that is uninhabited, so there are no humans or living things to be subjugated. However, colonialism was not only about the subjugation of subjects, but has also transformed landscapes, exploited resources and established narratives of the legitimate subordination of nature for profit. This has implications for how humans will develop their space activities and, similarly, runs the risk of establishing an unequal interplanetary system, in which only a few states and corporations will benefit. In light of a 'new space' economy where state actors are increasingly relying on private actors' technological innovation and support, it also becomes unlikely that social or environmental objectives will be prioritized over market objectives (Weinzierl 2018).

From a critical geopolitics perspective, this type of thinking manifests geopolitical knowledge and strategies of outer space, and is particularly problematic as it neglects any kind of environmental questions. The environment of outer space features only in broader considerations of the 'astropolitical environment' that may influence the possibility of establishing space dominance. When Dolman, for instance, speaks about space as 'a rich vista of gravitational mountains and valleys, oceans and rivers of resources and energy' (Dolman 2002: 61), these natures are not part of the considerations of the ecology of the space environment; instead, they are part of geopolitical thinking and the possibility of achieving dominance through control of strategic geographic areas in outer space. Similarly, the focus on different outer space environments such as low, medium and high Earth and geosynchronous orbit is driven by finding the most favourable spots for strategic satellites that would establish geostrategic advantage over Earth (Deudney 2020). Environmental questions only feature in terms of threats to national security or economic activities (for example, the destruction of communication and surveillance systems through space debris). With this rationale, not only are geopolitics extended to space, but too are the security logics that are often applied to environmental issues on Earth. However, including environmental questions would not only point to environmental pollution on Earth and in space, but would also necessarily ask what kind of injustices are imbricated in geopolitical space racing (Klinger 2021a, 2021b). This would allow for alternative conceptions of sovereignty, such as those that underpin ontologies of outer space as a global commons instead of geopolitics.

Outer space as a global commons

In contrast to the realist ontology of outer space as geopolitics stands a liberal ontology that understands outer space as a global commons. A territory is usually defined as a global commons when it does not fall under national jurisdiction and every state has access to it. Antarctica and the oceans are well-known examples of global commons, governed by the principle of the common heritage of humankind, and part of a broader liberal tradition to govern global environmental issues through cooperative agreements. International treaties like the United Nations Convention on the Law of the Sea (UNCLOS) and the Antarctic Treaty System (ATS) ensure that the different ecosystems are sustainably managed for mutual advantage and that environmental protection is prioritized over economic development objectives. For most legal scholars, outer space is another area where the principle of the global commons applies and in which historically a surprising degree of international cooperation has been shown, despite realist assessments of outer space as a site of great power competition (Cross 2021; Patton 2022). From a liberal perspective that conceptualizes outer space as a global commons, it then becomes a pertinent question to ask how governance can protect space from the tragedy of the commons (Patton 2022).

Governance relating to outer space has evolved in conjunction with the expansion of space activities. Ever since the advent of the first satellite when Sputnik entered outer space in 1957 as the first object from Earth, the need for regulation of scientific and military activities in outer space, as well as the entrance and re-entry of Anthropocentric objects, has become an important international political discussion. This led to the creation of an international organization dedicated to address these questions under the umbrella of the United Nations (UN) system, the UN Committee on the Peaceful Uses of Outer Space (COPUOS) in 1959. Today's governance of outer space is still largely based on those international treaties that were established in the early days of space exploration during the Cold War. Foundational in this respect is the Treaty on Principles Governing the Activities of States in the Exploration and Use of Outer Space, including the Moon and Other Celestial Bodies (Outer Space Treaty – OST) of 1967, to which 115 countries are parties and which all major space powers, such as the US, Russia and China, have ratified. The OST is regarded as setting the normative foundation of space activities when it states that outer space should serve only peaceful purposes and that it 'shall be free for exploration and use by all States' and for the 'benefit of all peoples' (UN 1967). With this purpose, the treaty aims to limit the possibilities for conflict in outer space by setting limitations to state sovereignty claims and appropriation in outer space. As Article II clearly states: 'Outer Space, including the moon and other celestial bodies, is not subject to national appropriation by claim of sovereignty, by means of use or occupation, or by any other means' (UN 1967, Article II). The Moon Agreement of 1979 was even more ambitious in attempting to safeguard outer space from any form of appropriation. It builds on the OST when it states that the Moon cannot be 'subject to national appropriation by claim of sovereignty, by means of use or occupation, or by any other means', but goes even further by calling the Moon and its resources 'the common heritage of all mankind' (UN 1979), alluding to the global commons principle. However, the Moon Treaty has only been ratified by 17 countries, none of which are the major space-faring states, and is therefore often described as a failed agreement.

An understanding of outer space as a global commons depicts an ontology which sees outer space as a space that belongs to everyone. This signals stewardship to protect outer space as a resource for all nations and to balance conservation and development which would similarly hinder private investments and the use of space resources. From the perspective of this liberal ontology, the OST (and the Moon Treaty) apply Westphalian thinking of territorially bounded sovereignty to outer space when demarcating the boundary between Earth and outer space, while similarly protecting outer space as a global commons by keeping states from claiming any sovereignty in space. In contrast to realist astropolitics, this reflects an optimistic outlook on the ability of such a regime to keep imperial and expansionary state politics in check. Deudney (2020) has argued that these governance regimes make an important contribution to interstate cooperation to collectively protect Earth from threats in outer space such as space weapons or asteroids, leading to increased planetary security.

In light of the recent rush to secure access to outer space and its resources by private and state actors alike, it comes as no surprise that this assessment is increasingly challenged. Hertzfeld, Weeden and Johnson (2015: 3) argue that the freedom to access space in Article I of the OST does not mean the actual physical space such as celestial bodies, but the activity to explore and use outer space (as the full treaty name of the OST also indicates). Arguing that there is an ambiguity about the appropriation principle in the text of the agreement (Pic, Evoy and Morin 2023), there are ongoing efforts to create competing regimes that would allow for the national and private appropriation of space resources – for instance, the Trump administration argued in 2020 that 'the United States does not view [outer space] as a global commons' (Executive Office of the President 2020). This assessment allowed for alternative international agreements that would reflect this understanding: the National Aeronautics and Space Administration (NASA) and the US Department of State followed up with the drafting of the Artemis Accords in the same year, which are based on the OST, but propose a specific interpretation of the treaty by allowing states to claim resources in outer space. The Artemis Accords have been signed in a short time by 53 countries (as of March 2025), suggesting that many countries are willing to re-interpret the rules set by the OST and allow private-sector commercial activities. This will have important repercussions for sovereignty in outer space. It would move away from a liberal understanding that outer space regimes can safeguard the security of sovereign states on Earth through peaceful cooperation in outer space towards a realist understanding that would allow states and private entities to partially extend their sovereignty to outer space.

Yet, where does the sovereignty of states begin and end in the first place? Constructivist scholars have long argued that the demarcations of territory are socially constructed and that nations and communities are imagined (Anderson 1983). Although constructed, these identities and ideas of belonging have material consequences as borders cannot only be drawn on maps, but can also be secured with fences, experienced through visa regimes and performed in border controls (Beaumont and Glaab 2023). In the case of outer space, the constructivist argument is even more compelling as the end of the territoriality of states and even planetary sovereignty can only be imagined. This is because there are geophysical limitations to the practice of sovereignty in outer space, which stand in contrast to the everyday practices of sovereignty on Earth. While all Anthropocentric objects from satellites over probes to spacecrafts need to be registered according to the UN Convention on Registration of Objects Launched into Outer Space to track the 'crossing' of the invisible border to outer space, sovereignty claims cannot be easily demarcated as a territory in outer space. Therefore, sovereignty claims by state (and nonstate) actors pertaining to outer space remain largely discursive. Practices of sovereignty have a rather exemplary symbolic character through, for instance, the famous flag-planting on the Moon or leaving behind of exploratory rovers and artefacts from scientific missions that remind future space missions that a territory is no longer *terra nullius*. The difficulties of defining, let alone claiming, sovereignty have led some scholars to the assessment that space cannot be regarded as a global commons as 'there is no sovereignty in space, as the edges of space are not defined' (Hertzfeld, Weeden and Johnson 2016: 20).

Despite the difficulty of demarcating boundaries between Earth and outer space, scholars try to define it by drawing on scientific explanations of nature. Critical geographers have suggested thinking about outer space as constructed and a sphere of the social (MacDonald 2007; Beery 2016), critically interrogating the binary categories of human/nature and Earth/outer space which are underlying these scientific arguments. Beery (2016) argues that outer space is socionaturally (that is, at the intersection between society and nature) constructed as a global commons and resource beyond sovereign territories. By examining the legal history of outer space, he demonstrates that early discussions on the governance of outer space pointed to the physical movement and spinning of Earth to show that it is not possible to simply extend sovereign borders vertically into outer space and to distinguish 'the space of outer space from terrestrial space' (Beery 2016: 95). Yet, despite this clear distinction between 'natures under state sovereignty (terrestrial space) from natures that are to be used for the benefit of humanity and accessible to all states (outer space)' (Beery 2016: 96), the OST lacks clarification in terms of where the boundary is located. The border between Earth and outer space (and the end of terrestrial sovereignty) is often defined as located beyond Earth's atmosphere (Williamson 2006: 27), and many including the International Astronautical Federation suggest it should be the so-called Kármán line at 80–100 kilometres above mean sea level, where the centrifugal force becomes dominant over aerodynamic lift and an aircraft 'becomes' a spacecraft (Lawrence et al 2022). While these discussions on boundaries seem to be part of a legal urge to arrive at definitional clarity, Beery reminds us that these constructions are situated in political, economic and social contexts with potentially uneven outcomes for different actors. A construction of outer space as a global commons is not necessarily a positive liberal vision of outer space governance in an equal and cooperative manner; instead, the definition of the nature of outer space can similarly lead to uneven benefits that favour states already actively involved in space activities (Beery 2016: 99).

Post-planetary politics and outer space governance beyond sovereignty

Realist and liberal ontologies of outer space bring forward a particular logic for approaching outer space governance. While a realist ontology shapes a worldview that takes state boundaries as given and sees possibilities to extend these to outer space in pursuit of national interests on Earth, a liberal ontology depicts a worldview of outer space as a global commons with shared responsibility and stewardship of the resources in outer space. The discussion of realist and liberal perspectives has similarly shown that nature has played an important role in scholarship on outer space politics, albeit with very different constructions of nature and consequently with different assessments of its value when these approaches assume that there is a geography and nature specific to the site of outer space. Both ontologies also rely on clear separations between Earth and outer space which are based on demarcations of nature built on natural science knowledge. As a result, realist as well as liberal ontologies take the constructed boundaries between planet Earth and outer space as given, while prioritizing the Anthropocentric view in this binary. In this way, they do not consider how these areas are interrelated or whether outer space can be regarded as an environment and ecosystem by itself. With increasing focus on the development of the space sector, its growing economic relevance, and more scholars and policy makers 'looking up', the sustainability of space activities and questions of nature have only recently come to the fore – mirroring similar discussions around the environment as a niche issue in IR (see Chapter 1). But an increasing number of scholars from different disciplinary fields have begun foregrounding the role of nature in their assessment of space activities.

Building on the existing legal framework, scholars point out that the OST identifies Earth orbit as an environment worth preserving, while related agreements such as the Space Liability Convention of 1972 make it possible to extend 'the concept of environment ... to orbital space' (Lawrence et al 2022). Showing the possible sustainability implications of new satellite constellations in outer space on astronomy through light pollution, potential collisions and debris of inactive satellites, as well as the effect on animal and plant ecosystems on Earth, scholars argue that orbital space should be considered part of the human environment and are calling for an ethical and sustainable approach to space (Lawrence et al 2022). In particular, Indigenous Peoples have highlighted the cultural significance of the sky. For instance, the Navajo Nation complained to NASA about plans to send cremated human remains to the Moon, citing the Moon's sacred place in Indigenous cultures and cosmology. Indigenous concerns are also based on the experience of forceful exclusion from Indigenous lands and the concern of the possible replication of these colonial histories in outer space (Milligan 2023). Also, the experiences of environmental injustices concerning the development of space infrastructures on Earth play a role, such as the social injustices that developed around the European Space Agency launch site in French Guiana (Redfield 2000). Others fear that the fast-growing space economy will lead to resource depletion in the future and argue for a precautionary approach that would reserve parts of the solar system as wilderness in the form of designated 'planetary parks' without options for human development (Elvis and Milligan 2019). Scholars in the field of sustainability research have developed the concept of 'Earth-space sustainability' which contends that sustainability concerns do not stop with imagined planetary boundaries, but that development on Earth is shaped by space activities and vice versa (Yap and Truffer 2022). Hence, they call for a more just and sustainable multi-planetary governance where 'environmental responsibility, ethics, and 'sense of place' extend beyond Earth' (Yap and Kim 2023: 4).

How can these recent analyses from disparate disciplinary perspectives – in legal studies, political science, geography or anthropology – on the role of nature for outer space governance be tied together? The intervention of a growing number of scholars in the field of global environmental politics that have proposed a 'planetary politics' provides a good point of departure. This literature has challenged us to understand politics beyond the traditional understanding of the nation state, but to acknowledge that a 'planetary politics' is necessary to address the environmental and sociopolitical challenges that have arisen as a result of the rapid modernization and globalization of capitalism (Burke et al 2016; Chakrabarty 2021; Marsili 2021). Acknowledging that humanity has become a geophysical force affecting all of Earth's systems (Litfin 2003), the planetary connects the biophysical world with human activities and shows the implications for planetary governance (Blake and Gilman 2024). Litfin's (2003) notion of planetary politics suggests the necessity of a holistic understanding of the interrelated processes that make up the entire planet, adding to recent calls to rethink IR's categories and the binary division between nature and humans (Dalby 2020; see Chapter 1).

Two contributions from this literature are particularly relevant to the argument provided in this chapter: For one thing, planetary politics makes nature the starting point for any discussions on governance by foregrounding how humans are part of Earth's ecosystems. This analytical starting point has normative implications and suggests that different species, human and nonhuman life must be taken care of equally. Hence, Burke et al (2016) advocate that humans should be 'Earth-worldly' and acknowledge their entanglement with other co-constituents of their worlds. For another thing, the planetary politics literature has encouraged IR scholars to think beyond the 'global' and in terms of the 'planetary'. They highlight the interplay between scales, pointing out that climate change is, for instance, similarly the result of local actions and the global economy, but with planetary implications (Litfin 2003). This also implies a move away from sovereign nation states as the main political authorities to solve such issues in their national jurisdictions or multilaterally and the need for planetary governmentality instead (Blake and Gilman 2024). Critical scholars of outer space politics would likely concur, but add the crucial observation that 'planetary politics does not end at the edge of the Earth's atmosphere' (van Wingerden and Vigneswaran 2024: 601). Accordingly, a *post-planetary* politics would go one step further and acknowledge that the natures of Earth and outer space are entangled. This does not only apply in terms of the growing space activities by private and public actors in outer space, but also in terms of the use of space-based technologies on Earth and how outer space shapes activities on Earth in terms of governance and commercial activities (Klinger 2021b).

In other words, a post-planetary ontology of outer space requires a gestalt switch from an Anthropocentric perspective of 'looking out' from Earth – as realist and liberal ontologies of outer space do – to 'looking in'. The first picture from outer space, the 'Earthrise' photograph taken in 1968 during the Apollo 8 mission, showed for the first time the view of Earth from outer space. This was the beginning of the so-called Earthrise era and brought about an 'earthly vision' (Lazier 2011), which catalysed an environmental consciousness and movement for the preservation of the planet in awe of the finiteness of Earth (Lekan 2014). It also changed perspectives between Earth and outer space for the first time. Realist and liberal perspectives alike base their understanding of sovereignty on an Anthropocentric view that is looking out from Earth. By applying Westphalian conceptions of terrestrial sovereignty to outer space, they aim to demarcate the boundaries between Earth and outer space as well as strategic territories in outer space itself. The Earthrise photograph reversed that kind of thinking by looking in from outer space. It showed not only the place of Earth within the multi-planetary universe but also made visible the artificial construction of state boundaries. After all, from space there are no boundaries visible on Earth; there is only nature. Instead of territorial borders that are drawn on political maps of the Earth, there are oceans, forests, mountains and deserts visible that disregard the terrestrial construction of sovereign states. From that perspective, 'Earth itself becomes a singular locality' (Messeri 2016: 11). A post-planetary relational ontology thus challenges the hierarchies of scales and goes beyond a multiscalar approach as it not only adds outer space to the local, national and global levels of analysis, but also challenges the very existence of such levels. When Earth is in planetary relations with other spaces (Olson and Messeri 2015), a post-planetary politics would endorse an understanding of sovereignty that frees itself from 'the terrestrial trap' (van Wingerden and Vigneswaran 2024).

Conclusion: governing outer space as nature

This chapter has explored how realist, liberal and what I called post-planetary ontologies of outer space imply radically different ways of relating to the relationship of Earth and outer space, as well as humans and nature. Importantly, these ontologies undergird political discourses and strategies, and are constitutive of different possible governance arrangements of outer space. From the post-planetary ontology advanced here, the spaces of outer space and planet Earth are interrelated, making outer space not external but integral to environmental politics on Earth. Indeed, recognition of outer space as an environment or ecosystem collapses the binary division of humans and nature as well as Earth and space (or even human and nonhuman). A questioning of 'the terrestrial grounding of concepts' (Olson and Messeri 2015: 38) also leads us to question the applicability, extension and transformation of what has traditionally been labelled as sovereignty politics in IR. Terrestrial sovereignty cannot simply be extended to outer space and sovereignty games reach their limits when there are no sovereigns involved.

Different ontologies of outer space will necessarily leave us with different opportunities to govern outer space and regulate technological innovation. However, the development of space technologies and activities and their governance are not natural and path-dependent, but are made through decisions. So far, the OST remains universal in its approach to safeguarding outer space from appropriation, but there are increasing efforts to identify ambiguity that would allow, for instance, for exclusive economic zones and the privatization of outer space. In contrast, adopting a relational ontology and acknowledging the co-constitution of Earth-space would allow for stewardship and careful development of space activities, and would provide a critical stance towards outer space as a new terrain to pursue states' geopolitical strategies on Earth. Post-planetary politics also has ethical repercussions, prompting questions and discussions to address historical and future injustices in space exploration. Will humankind be able to do better and avoid the mistakes of its colonial past? Can the space endeavour be democratized to ensure that not only the powerful but also a variety of countries and actors have access to space? Should celestial bodies have rights as nonhuman legal persons? This chapter has shown that sovereignty claims serve different political goals and how we relate to outer space matters for its sustainable governance. Taking outer space seriously as nature will also shape the future of planet Earth in a multi-planetary world.

Acknowledgements

I extend my heartfelt thanks to the editors for their unwavering support and invaluable feedback on the chapter. Our discussions have been crucial in shaping its final version. The research and writing of this chapter was supported by the NordSpace project (Nordic Space Infrastructures: Environment, Security, and Future Imaginaries of Outer Space in the High North, project # 343446), funded by the Norwegian Research Council.

References

Anderson, B. (1983) *Imagined Communities: Reflections on the Origin and Spread of Nationalism*, revised edn. New York: Verso.

Beaumont, P. and Glaab, K. (2023) 'Everyday Migration Hierarchies: Negotiating the EU's Visa Regime', *International Relations*, 00471178 231205408.

Beery, J. (2016) 'Unearthing Global Natures', *Political Geography*, 55, 92–101.

Blake, J.S. and Gilman, N. (2024) *Children of a Modest Star: Planetary Thinking for an Age of Crises*. Stanford: Stanford University Press.

Burke, A. et al (2016) 'Planet Politics: A Manifesto from the End of IR', *Millennium: Journal of International Studies*, 44(3), 499–523.

Chakrabarty, D. (2021) *The Climate of History in a Planetary Age*. Chicago: University of Chicago Press.

Cross, M.K.D. (2021) 'Outer Space and the Idea of the Global Commons', *International Relations*, 35(3), 384–402.

Dalby, S. (2020) *Anthropocene Geopolitics: Globalization, Security, Sustainability*. Ottawa, Ontario: University of Ottawa Press.

Deudney, D. (2020) *Dark Skies: Space, Expansionism, Planetary Geopolitics, and the Ends of Humanity*. New York: Oxford University Press.

Dolman, E.C. (2002) *Astropolitik: Classical Geopolitics in the Space Age*. London: Frank Cass.

Duvall, R. and Havercroft, J. (2008) 'Taking Sovereignty out of This World: Space Weapons and Empire of the Future', *Review of International Studies*. 34(4), 755–775.

Elvis, M. and Milligan, T. (2019) 'How Much of the Solar System Should We Leave as Wilderness?', *Astra Astronautica*, 162, 574–580.

Executive Office of the President (2020) *Executive Order 13914 of April 6, 2020: Encouraging International Support for the Recovery and Use of Space Resources*, 85 Fed. Reg. 20381.

Havercroft, J. and Duvall, R. (2009) 'Critical Astropolitics: The Geopolitics of Space Control and the Transformation of State Sovereignty', in *Securing Outer Space*. Abingdon: Routledge, pp 42–58.

Hertzfeld, H.R., Weeden, B. and Johnson, C.D. (2015) 'How Simple Terms Mislead Us: The Pitfalls of Thinking about Outer Space as a Commons'. *International Astronautical Congress*, 15.

Hertzfeld, H.R., Weeden, B. and Johnson, C.D. (2016) 'Outer Space: Ungoverned or Lacking Effective Governance? New Approaches to Managing Human Activities in Space', *SAIS Review of International Affairs*, 36(2), 15–28.

Klinger, J.M. (2021a) 'Critical Geopolitics of Outer Space', *Geopolitics*, 26(3), 661–665.

Klinger, J.M. (2021b) 'Environmental Geopolitics and Outer Space', *Geopolitics*, 26(3), 666–703.

Lawrence, A. et al (2022) 'The Case for Space Environmentalism', *Nature Astronomy*, 6(4), 428–435.

Lazier, B. (2011) 'Earthrise; or, the Globalization of the World Picture', *American Historical Review*, 116(3), 602–630.

Lekan, T.M. (2014) 'Fractal Earth: Visualizing the Global Environment in the Anthropocene', *Environmental Humanities*, 5(1), 171–201.

Lerner, A.B. and O'Loughlin, B. (2023) 'Strategic Ontologies: Narrative and Meso-Level Theorizing in International Politics', *International Studies Quarterly*, 67(3): sqad058.

Litfin, K.T. (2003) 'Planetary Politics', in J. Agnew, K. Mitchell, and G. Toal (eds) *A Companion to Political Geography*. Chichester: Wiley, pp 470–482.

MacDonald, F. (2007) 'Anti-Astropolitik: Outer Space and the Orbit of Geography', *Progress in Human Geography*, 31(5), 592–615.

Marsili, L. (2021) *Planetary Politics: A Manifesto*. Cambridge: Polity Press.

Messeri, L. (2016) *Placing Outer Space: An Earthly Ethnography of Other Worlds*. Durham: Duke University Press.

Milligan, T. (2023) 'From the Sky to the Ground: Indigenous Peoples in an Age of Space Expansion', *Space Policy*, 63: 101520.

Morin, J. and Richard, B. (2021) 'Astro-environmentalism: Towards a Polycentric Governance of Space Debris', *Global Policy*, 12(4), 568–573.

Musgrave, P. and Nexon, D.H. (2018) 'Defending Hierarchy from the Moon to the Indian Ocean', *International Organization*, 72(3), 591–626.

Olson, V. and Messeri, L. (2015) 'Beyond the Anthropocene: Un-Earthing an Epoch', *Environment and Society*, 6(1), 28–47.

Patton, D. (2022) *Is Space a Global Commons?* Secure World Foundation, 1–36. Available at: https://swfound.org/media/207517/swf_brief_is_space_a_global_commons_pp2301_final.pdf

Pic, P., Evoy, P. and Morin, J.-F. (2023) 'Outer Space as a Global Commons: An Empirical Study of Space Arrangements', *International Journal of the Commons*, 17(1), 288–301.

Redfield, P. (2000) *Space in the Tropics: From Convicts to Rockets in French Guiana*. Berkeley: University of California Press.

Redfield, P. (2002) 'The Half-Life of Empire in Outer Space', *Social Studies of Science*, 32(5–6), 791–825.

United Nations (1967) 'Treaty on Principles Governing the Activities of States in the Exploration and Use of Outer Space, Including the Moon and Other Celestial Bodies (Outer Space Treaty)'.

United Nations (1979) 'Agreement Governing the Activities of States on the Moon and Other Celestial Bodies (Moon Agreement)'.

Van Wingerden, E. and Vigneswaran, D. (2024) 'The Terrestrial Trap: International Relations beyond Earth', *Review of International Studies*, 50(3), 600–618.

Weinzierl, M. (2018) 'Space, the Final Economic Frontier', *Journal of Economic Perspectives*, 32(2), 173–192.

Williamson, M. (2006) *Space: The Fragile Frontier*. Reston: AIAA.

Yap, X.-S. and Kim, R.E. (2023) 'Towards Earth-Space Governance in a Multi-planetary Era', *Earth System Governance*, 16: 100173.

Yap, X.-S. and Truffer, B. (2022) 'Contouring "Earth-Space Sustainability"', *Environmental Innovation and Societal Transitions*, 44, 185–193.

7

World (Re)Ordering through Green Growth and Degrowth Futures

Bruna Bosi-Moreira and Matthias Kranke

Introduction

The future is always now. Imaginaries of certain futures, rather than others, affect the contemporary realm of possibility, as neatly captured in the term 'present future'. According to Niklas Luhmann (1976: 140), a genuine present future is not predetermined, but leaves ample 'room for several mutually exclusive future presents'. This idea has been picked up and specified in recent interdisciplinary work on how anticipation creates 'environmental futures' (Pavlínek and Petr 2004; Granjou et al 2017), 'Anthropocene futures' (Berkhout 2014; Erickson 2020) or, more specifically, 'climate futures' (Vervoort and Gupta 2018; Bhavnani et al 2022). In these accounts, too, imagined future states of the (global) environment reach 'back' into the present to direct efforts to protect and repair increasingly stressed or already damaged ecosystems. In terms of the volume's analytical framework, global sustainability governance constructs environmental issues – or 'nature' as an object – with reference to not only their present constitution but also their anticipated future constitution.

In this chapter, we extend this line of work by exploring the environmental futures underpinning 'green growth' and 'degrowth' (or, more broadly, 'postgrowth') agendas. As 'ecopolitical projects' (Buch-Hansen and Carstensen 2021), both green growth and degrowth have surged in popularity in recent years as potential responses to the global environmental crisis facing humanity. Proponents of green growth and advocates of degrowth agree that the current fossil fuel-based order is unsustainable, which means that it must ultimately be overcome. Yet at the core of each of these two political programmes are very distinct imaginaries about the future state of planetary ecosystems and their interplay with contemporary societies, which bear differently on the problems identified and the solutions offered in the present. Notwithstanding recent juxtapositions of green growth and degrowth/postgrowth (see Buch-Hansen and Carstensen 2021; Hasselbalch et al 2023; Polewsky et al 2024), how each side imagines future environmental issues has remained a relative blindspot in postgrowth scholarship. Conversely, analyses of environmental futures have tended to overlook green growth and degrowth as two major stances in the global sustainability debate that operate through imaginaries of future possibilities and constraints.

More specifically, we argue that both strands largely fail to engage questions of world ordering encapsulated in their own analyses and recommendations. After all, the polarized debate between green growth and degrowth proponents is not merely about who offers the better toolkit for comprehending and addressing present and future sustainability problems. Behind such claims stand particular understandings of how the world ought to be reordered in the sustainability transition. Despite recognizing the global scale of the problem and the need for sustainability measures, green growth and degrowth futures more often diverge than converge. As we demonstrate through a comparative illustrative analysis of a few documents, green growth and degrowth futures convey different ideas about world order and, in turn, the extent and type of reordering required for sustainability. In other words, each perspective constructs future nature in ways that imply a particular approach to governing present nature.

The chapter proceeds as follows. We open with a review of the literature that links research on environmental (and related) futures with the notion of world order(ing). This conceptual foundation informs our comparative analysis of a small selection of key documents, which shows that green growth and degrowth futures embrace different understandings of world order. We then reflect on what these insights entail for our grasp and use of key International Relations (IR) concepts in the wider context of global sustainability governance. In the conclusion, we summarize why 'nature's governance reversed', as the editors of the collection call it, must also be understood in temporal terms – in our case, as imaginaries that invite us to see sustainability refracted through different lenses onto the future.

Environmental futures and world (re)ordering

Recent years have seen increasing academic interest in futures as 'forward-looking imaginaries' that define present possibilities across a variety of transnational domains (Berten and Kranke 2024: 614; see also Berten and Kranke 2022). Some such futures have an explicitly environmental character in delineating potential trajectories of human-nature or society-ecology relations broadly conceived. Environmental futures can be described as forward-looking imaginaries that sketch 'the political horizon of anthropogenic ecological risks on a planetary scale' (Granjou et al 2017: 5). Actors inevitably need to make choices about how to translate an inherently uncertain and unknowable future into a tangible outlook (de Jouvenel 2012: 5). Importantly, environmental futures entail imaginaries of world order, although these elements may in many cases be more implicit than explicit.

The tensions surrounding the present liberal international order have been debated across the IR discipline. Many contributions have offered particular takes on their causes and consequences (Ikenberry 2018; Eilstrup-Sangiovanni and Hofmann 2020; Lake et al 2021). Ecological change also factors into that equation, both weakening the existing world order (Werrell and Femia 2016) and being exacerbated by the resulting disorder (Eckersley 2023: 107). Thus, green growth and degrowth proposals can be understood, on the one hand, as outcomes born out of a declining world order and, on the other hand, as attempts of world (re)ordering in the face of anthropogenic degradation of the biosphere. There is widespread agreement that global ecological change has already begun to and will increasingly continue to undermine world order through its socioecological impacts. World disorder becomes more likely as temperatures rise further and especially when critical tipping points of the Earth system are reached. In short, we consider green growth and degrowth as approaches to the future that imply different world (re)ordering efforts in response to intersubjectively shared perceptions of disorder. Figure 7.1 provides a sketch of this foundational assumption for the arguments advanced here.

Our perspective aligns with the broad conception of order applied to this volume. At the same time, it follows Eckersley's (2023: 106, emphasis in original) argument that '[p]olitical orders ... are necessarily always *socio-ecological* orders because they interact with ecological communities to co-produce different local, regional and world "ecologies" or "natures".' It is no coincidence that others have characterized green growth and degrowth as 'ecopolitical projects' rooted in diverging paradigmatic diagnoses and prescriptions (Buch-Hansen and Carstensen 2021). If we thus understand order as a manifestation of rules about human-human and human-nature

Figure 7.1: Environmental futures and (dis)order

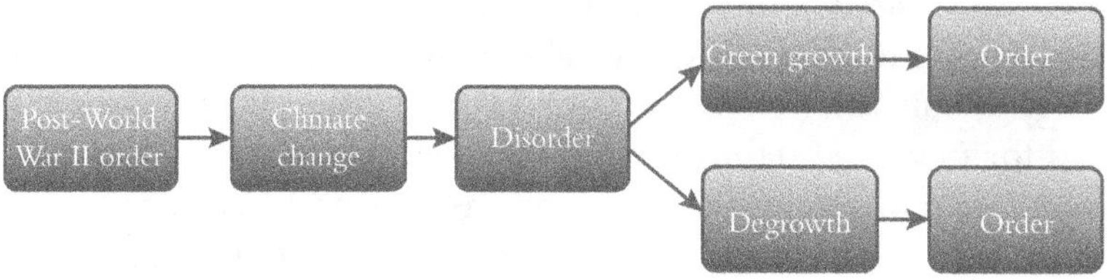

Post-World War II order → Climate change → Disorder → Green growth → Order
Disorder → Degrowth → Order

Source: Authors

relations that imply certain rights and certain responsibilities, green growth and degrowth not only feast on distinct policy menus but also embrace different world (re)ordering principles.

While the literature has widely demonstrated how actors produce, validate and disseminate environmental futures in global politics, we know much less about their underlying ideas of world (re)ordering. The work of the Intergovernmental Panel on Climate Change (IPCC), for example, revolves around specifying potential futures of global warming (Beck and Mahony 2018). When the IPCC develops climate change scenarios, they are based on certain assumptions about what humans and nonhumans will do within changed or largely unchanged sociotechnical systems and ecosystems. Similarly, forest scientists grapple with the potential evolution of forests within particular '"micro-regimes" of anticipation' that combine ontological, epistemological and methodological resources (Dolez et al 2019). Different assumptions underpin these frameworks so that forests are assigned different future roles in the wider set of human-nature relations. To give a final example, when a country seeks to join a supranational entity, such as the European Union (EU), new environmental futures are developed and come to the fore (Pavlínek and Petr 2004). Again, these futures rely on a particular understanding of world order, with the country being a new member of a supranational polity governed by specific rules in the environmental realm and beyond. More fundamentally, decision-making processes are often directed towards such futures (Gibbs and Flotemersch 2019), which can, in turn, stabilize or disrupt the existing political order.

In the next section, we look more closely at green growth and degrowth futures as concrete manifestations of environmental futures. Given the chasm between the two camps, we can expect them to also rally around different conceptions of how humans should engage with the biosphere at a time of escalating planetary crisis. As we show, they diverge quite sharply in their assessments of the impact of humans' economic activities on ecosystems.

Green growth and degrowth as world (re)ordering approaches

The literature that examines the green growth and degrowth/postgrowth paradigms and policy agendas (for example, Cosme et al 2017; Buch-Hansen and Carstensen 2021; Hasselbalch et al 2023; Polewsky et al 2024) provides us with some clues about what kind of future imaginaries each community may embrace. Put bluntly, green growth proponents see a need for measured reforms while degrowth advocates aspire to deeper changes, often including a deliberate move 'beyond capitalism' (Schmelzer et al 2022). Buch-Hansen and Carstensen (2021) capture this important difference by associating the green growth agenda with 'third-order change' and the degrowth agenda with 'fourth-order change', thus extending Hall's (1993) classic paradigm typology. Informed by such insights, we first formulate our theoretical expectations in this section before turning to a few illustrative examples that can be said to represent typical green growth and typical degrowth futures.

Green growth proponents usually imagine the future as an evolutionary extension of the political, economic and social status quo. Indeed, they attempt to 'green' the status quo – and thus restore political order – by relying more on the expected benefits from technological innovations than on the effects of social innovations. Their silver bullet for such a technologically induced transformation is the realization of market-driven efficiency gains, which serves as an explicit or implicit justification for why the current system, including the institutionalized growth imperative, can be maintained. In more technical terms, this silver bullet is called 'decoupling', whereby the ratio between a technology's benefits and its resource consumption (for both inputs and outputs) improves (Fücks 2014; Stoknes and Rockström 2018). In this line of thinking, the global disorder wrought about by already occurring and expected further ecological change gets resolved through a measured approach that combines technological solutions at scale with economic growth rendered sustainable (for example, Pollin 2018). Such a position can be linked with global distributional concerns, such as when green growth is reserved as a policy priority for materially less affluent countries (Okereke 2024).

By contrast, degrowth advocates stress that the status quo must be overcome swiftly to establish a new political order. Contrary to some mistaken claims, degrowthers do not reject technological innovations outright. Instead, they want *certain* – that is, socially and ecologically beneficial – technologies to be developed, rolled out and scaled up (Hickel 2024), while also insisting on complementary social innovations through different forms of social provisioning (Brand et al 2021). In other words, because efficiency gains alone will not suffice, they must be complemented by sufficiency strategies. Technologies must thus be able to achieve and sustain absolute, rather than just relative, forms of decoupling, especially given the danger of new resource demands resulting from 'rebound effects' (Haberl et al 2020; Hickel and Kallis 2020). From this perspective, green growth proposals appear not as solutions to but as extensions of disorder. Merely tweaking a political order based on rampant socioecological injustices would thus leave humanity on a path towards more biospheric degradation and more social upheaval, with the long-term spectre of human extinction looming large. While the idea of degrowth pre-dates the political mainstreaming of climate change and other global sustainability issues, recent degrowth proposals have been formulated in acute awareness of the many manifestations of the planetary crisis (for example, Asara et al 2015; Wiedmann et al 2020; Schmelzer et al

2022: 3–4). Not least for this reason, reordering as envisaged by degrowthers is both comprehensive and deep.

To move beyond hypotheses, we consulted a small number of sources that could yield some insights into the dominant environmental futures of the green growth and degrowth communities. These communities, particularly the green growth one, which has managed to mainstream its ideas, are sizeable so that many actors would be suitable candidates for analysis, from national governments (Purdey 2010) to international organizations (Meckling and Allan 2020); even the degrowth community seems to have gained followers in recent years, at least among activists and academics. However, our goal here is not to offer a comprehensive empirical analysis of a large number of actors on each side. Our more modest ambition is to take an illustrative snapshot of the environmental futures typical of each position.

To this end, we selected four texts as representations of green growth and degrowth imaginaries, respectively: on the green growth side, the 2011 report *Towards Green Growth* by the Organisation for Economic Co-operation and Development (OECD 2011), and the 2012 report *Inclusive Green Growth: The Pathway to Sustainable Development* by the World Bank (2012); on the degrowth side, the declaration formulated collectively by the participants of the first-ever Degrowth Conference in April 2008 ('Declaration' 2008) and the 'manifesto' written by the editorial collective of the new *Degrowth Journal* and published on its website in May 2023 ('The manifesto of Degrowth Journal' 2023). The OECD and the World Bank have been leading voices and early movers in favour of green growth on the international policy scene (see Meckling and Allan 2020), thus shaping the views of other actors. The Degrowth Conferences have become established regular get-togethers of the transnational degrowth community, initially on a biennial and now an annual basis, while the *Degrowth Journal* can be seen as its new intellectual mouthpiece. The following illustrations rely on a very selective reading of key statements from the two green growth reports (142 and 171 pages, respectively), focusing on introductory definitions and related explanations, and a full reading of the two short degrowth documents (two and five pages, respectively).

To begin, we can note that the OECD has often spearheaded the green growth movement in the policy world. Its activity in this respect has emerged from a double recognition: namely, first, that current economies are not sustainable; and, second, that economic growth continues to be central to the creation and maintenance of future societal welfare. As the OECD (2011: 18) puts it:

> Thus the world faces twin challenges: expanding economic opportunities for a growing global population; and addressing

environmental pressures that, if left unaddressed, could undermine our ability to seize these opportunities. Green growth is where these two challenges meet and about exploiting the opportunities which lie within.

The World Bank (2012: 6) largely echoes this sentiment with a view to its mission as an international development institution, arguing:

> Thus, growth is a necessary, legitimate, and appropriate pursuit for the developing world, but so is a clean and safe environment. Without ambitious policies, growth will continue to degrade the environment and deplete resources critical to the welfare of current and future generations.

So what is 'the political horizon of anthropogenic ecological risks on a planetary scale' (Granjou et al 2017: 5) behind these statements? For both organizations, a critical challenge lies in attenuating market inefficiencies as much as possible because they lower economic growth, while reducing resource use (and hence environmental pressures), relatively speaking, for each unit produced or consumed (OECD 2011: 21–22; World Bank 2012: 3–4). Although this orientation clearly acknowledges human-made global ecological problems, it seeks to address them from within the existing economic system by promoting a combination of reformed policy frameworks and corrected market incentives that steer actors in a green(er) direction. As a result, both the OECD (2011: 10) and the World Bank (2012: 2) assume that new green technologies will deliver solutions, ultimately resolving the trade-off between economic growth and ecological sustainability. Larger systemic questions – for example, the viability of capitalism as such – are typically bracketed in green growth environmental futures. Changes are evolutionary and cumulative to enable a relatively smooth long-term transition, rather than presupposing system change. The future then looks bright in these imaginaries by virtue of innumerable small steps that will add up over time.

Degrowth environmental futures could hardly be more different, displaying a clear scepticism of green growth-style technocratic interventions to overcome unsustainable structures and practices. The second statement by the participants of the inaugural Degrowth Conference, which was convened in Paris, exemplifies this difference: 'Despite improvements in the ecological efficiency of the production and consumption of goods and services, global economic growth has resulted in increased extraction of natural resources and increased waste and emissions' ('Declaration' 2008: 1).

The editorial for the new *Degrowth Journal* understandably concentrates more on publishing practices and principles, but it still speaks to wider issues that reveal the sort of forward-looking imaginaries treasured within the degrowth community:

> Let us apply the logic of commoning and create a safe 'knowledge commons' in order to protect our ability to study important questions, starting with imagining life beyond growth and capitalism. The journal *Degrowth* embodies core values of degrowth such as autonomy and solidarity: its goal is to decommodify knowledge, to go against the mass profiteering of commercial publishers and defend free access to science for the common good. ('The manifesto of Degrowth Journal' 2023: 2, emphasis in original)

This statement dovetails with the idea of 'right-sizing' economies, which is a prominent idea in the Declaration. Rich economies with excessive resource demands are 'right-sized' by being re-aligned with biophysical dynamics, while poorer economies are 'right-sized' by being enabled to increase consumption so that decent living standards for all can be reached ('Declaration' 2008: 2). This rendering has clear affinities with what has in recent years been discussed under the label of 'doughnut economics', which combines ecological limits with minimum socioeconomic thresholds (Raworth 2017; for an application, see Fanning et al 2022). Degrowth environmental futures tend to foreground community planning while doubting that market incentives can be the primary triggers for truly transformative technological innovations. Frequently, the envisaged changes are presented as having to either begin from or amount to a deep structural change: the overturning of capitalism as the predominant economic system of our time, with all its distinct incentives and constraints. While the mainstreaming of nongrowth ideas and practices may already start under capitalism (Albert 2024), degrowthers mostly sketch futures in which capitalism is not simply more regulated, but eventually abandoned. The future then looks bright by virtue of a big (initial) step that makes further steps possible.

In sum, green growth and degrowth environmental futures differ markedly, although both aspire to harmonious human-nature relations on the basis of an intact and thriving biosphere. In light of the preceding discussion, one could provocatively suggest that the future looks bright for green growth advocates if degrowth imaginaries are shunned; similarly, it looks bright for degrowthers if prevailing (green) growth imaginaries are replaced. It is instructive to analytically locate these differences within the multilevel perspective (MLP), which has become popular in the sustainable transitions literature (Geels 2002, 2011). While green growth futures zoom in on the interactions between the micro- and the meso-levels – that is, the interplay between 'niches' and 'sociotechnical regimes' – degrowth futures draw far more attention to the need for macro-level or 'landscape' changes.

World order is most certainly a landscape phenomenon, which means that degrowth reordering attempts tend to lean towards systemic transformation, as evidenced by calls for an altogether different system. In the next section, we discuss how these insights could reshape the study of global politics.

Implications for the study of global politics

The distinct environmental futures envisioned by the green growth and degrowth approaches have different implications for key IR concepts, debates or areas of enquiry. Several aspects of IR theorizing and practice could be affected by the two approaches presented in this chapter. In this section, we focus on four key IR concepts – namely 'sovereignty', 'power', 'security' and 'justice' – to analyse how green growth and degrowth thinking could redirect the study of global politics (see also Hasselbalch and Kranke 2024).

Sovereignty is arguably the foundation stone of international relations. Green growth futures reinforce states' authority to define the course of development within their borders. Green New Deals, such as the European Green New Deal or the Inflation Reduction Act in the US, serve as good illustrations. Despite their efforts to begin phasing out old fossil fuel energy, the focus is distinctly national and statist (see Mastini et al 2021). By contrast, degrowth futures are typically much more transnational in orientation, being the product of activism by social movements and academics. Moreover, limiting economic growth to stay within planetary boundaries would impinge on states' traditionally accepted right to steer the economy towards continuous expansion as a basis for societal welfare (Purdey 2010; Hasselbalch et al 2023). In fact, many degrowthers are deeply sceptical of the state's role, often contending that the most affluent states have contributed the most to global unsustainability coupled with social deprivation in many other places (Dorninger et al 2021; Fanning et al 2022). One main reason is that states' environmental policies have been designed to fix local pollution, but not to work against the routine externalization of the effects of high levels of material consumption (Hausknost 2020: 26–27).

Like sovereignty, power is a core IR concept. Among its many dimensions, economic power has been perceived as essential, together with military power. Green growth futures tend to reproduce that logic as states are portrayed as unable – or ill-advised – to stop pursuing growth, given its contribution to societal stability (via certain forms of socioeconomic redistribution). By combining the imperative to grow with sustainability demands, this perspective then relies to a significant extent on technological solutions, which create new winners and losers. By contrast, degrowth futures challenge established notions of power, broadly defined, by outlining and actively recommending limits to states' ability to seek power through economic strength. Therefore, a degrowth future might not only impact the current global power imbalance by narrowing the economic gap between developed nations and the Global South, but also change the very understanding of what power in international relations means, stressing less tangible dimensions of power, such as power as authority instead of power as coercion (Hurd 1999; Lake 2013).

Another fundamental IR concept is security, which scholars have traditionally understood along interstate lines. In the context of green growth and degrowth futures, this understanding shifts somewhat towards questions of resource security and its socioecological implications. While green growth futures may not require secure access to fossil fuels in politically unstable regions, they are feasible only to the extent that large amounts of critical minerals are available to produce technologies for generating and storing low-carbon energy, such as solar panels, wind turbines or batteries. This scenario spawns other security issues, such as dependence on a few producers and the consequent risk of these materials being weaponized in the event of geopolitical conflicts. In addition, because it is such a strongly technology-based future, the race to dominate green technologies could lead to the (further) securitization of not only ecological change but also the technologies that green growth presents as solutions. Degrowth futures ease the resource burden, which logically means fewer disputes over their access and use. However, this environmental future still has security implications, especially if the underlying economic issues get securitized (Buzan et al 1998: 22). Following that logic, lower economic growth could ultimately weaken a state's military capabilities. In full accordance with degrowth principles, measures 'to scale down ecologically destructive and socially less necessary production' can focus on, among other things, weapons and other military equipment (Hickel 2021: 1108; see also Tipping Point North South 2023).

Finally, different environmental futures imply different understandings of the concept of justice in IR. Debates on environmental justice have already highlighted the uneven effects that environmental degradation has on different societal groups and countries (Temper et al 2018; Dengler and Seebacher 2019). For example, climate justice is often at the centre of deadlocks in climate-related negotiations around historical responsibility, and related duties for loss and damage payments. When it comes to the future, the underlying question is: green growth or degrowth *for whom?* As we have seen earlier, green growth futures are contradictory in this respect. While aiming to address the material needs of developing countries, green growth also creates justice-related problems. Considering that the critical minerals needed for green energy and mobility are highly concentrated in developing or least developed countries, it could reproduce neocolonial and unjust practices in what has been called 'green sacrifice zones' (Zografos and Robbins 2020). It is important to underline that degrowth futures do not simply promote lower levels of economic activity across the board.

Table 7.1: Environmental futures and IR concepts

Potential implications of the two environmental futures		
Key concept	Green growth	Degrowth
Sovereignty	State-led future	Transnational movement at the substate level
Power	Economic power central Power as resources	Less tangible dimensions of power Power as authority based on legitimacy
Security	Securitization of green technologies	Less demand for resources, so fewer disputes over them Military sector intentionally affected
Justice	Growth for all countries Reproduction of neocolonial and unjust extractivist practices	Differentiation between Global North and South Alternative conceptions of development

Source: Authors

Instead, sectors are distinguished on the basis of their positive and negative socioecological effects, rather than their contribution to gross domestic product (GDP) (Hickel 2021: 1108). Moreover, degrowth futures envision ways of narrowing the gap between the Global North and South because degrowth is a programme aimed at wealthy countries, as well as at wealthy people in lower-income countries. Such visions therefore also dialogue with IR debates on global inequalities, particularly arguments advanced by postdevelopment scholars (Escobar 2015; Ziai 2015; Lang 2024). As it detaches development from economic growth, especially Western models of growth, it could open up space for alternative understandings of economic, social and ecological justice.

In sum, thinking through the lens of green growth and degrowth futures has tangible implications for how we approach key IR concepts. Table 7.1 summarizes the potential ways in which the two broad types of environmental futures juxtaposed here could reshape such concepts and, ultimately, the study of global politics.

Conclusion

In this chapter, we have begun to connect two streams of work that have thus far remained rather marginal to the field of IR. While research on global futures, environmental or otherwise, has taken off in recent years, the wider community still displays little interest in relating the green growth versus degrowth debate to dynamics of global politics. We have brought together these so far separate conversations to explore how different imaginaries of green growth and degrowth invoke particular understandings of current world disorder, sketching certain avenues for future reordering. We have argued that, despite their clear differences, both green growth and degrowth futures start from the understanding that the world currently exhibits a certain degree of disorder. The two approaches can then be read as competing political offers about how to best restore world order, which derive from discrepant assessments of the effects of economic growth on the biosphere. Whereas green growth futures revolve around the promise of making growth sustainable via new, more efficient technologies, degrowth futures present such a scenario as implausible, instead promoting a new order in which humans exercise restraint vis-à-vis other species and ecosystems in general.

IR scholars do not need to get bogged down in the often quite binary green growth versus degrowth debate, which has occupied many environmental economists (typically being in the green growth camp) and ecological economists (typically being in the degrowth camp). But continuing to downplay a fundamental lesson from the debate, namely that the politics of world (dis)order involves the biosphere quite directly (see again Eckersley 2023), risks leaving the field incapable of contributing literally vital insights in incredibly testing times (Burke et al 2016; Harrington 2016; Newell 2024). The debate on green growth and degrowth, heated though it is at times, serves as a reminder of the wider relevance of environmental futures: they do not 'merely' reimagine the place and role of humans and various nonhumans, as well as the ecosystems in which all of them are ultimately embedded; these futures also reimagine global configurations of order and disorder.

Acknowledgements

The authors contributed equally to this chapter. We wish to express our gratitude to the volume editors, Elana Wilson Rowe, Paul Beaumont and Lucas de Oliveira Paes, for their support and guidance. Matthias Kranke gratefully acknowledges funding from the Eva Mayr-Stihl Foundation for a one-year Young Academy for Sustainability Research Fellowship at the Freiburg Institute for Advanced Studies (FRIAS), University of Freiburg, Germany (from October 2023 to September 2024), during which the chapter was developed.

References

Albert, Michael J. (2024) 'Growth hegemony and post-growth futures: a complex hegemony approach', *Review of International Studies*, 50(5): 932–942.

Asara, Viviana et al (2015) 'Socially sustainable degrowth as a social–ecological transformation: repoliticizing sustainability', *Sustainability Science*, 10(3): 375–384.

Beck, Silke and Martin Mahony (2018) 'The politics of anticipation: the IPCC and the negative emissions technologies experience', *Global Sustainability*, 1, e8: 1–8.

Berkhout, Frans (2014) 'Anthropocene futures', *Anthropocene Review*, 1(2): 154–159.

Berten, John and Matthias Kranke (2022) 'Anticipatory global governance: international organisations and the politics of the future', *Global Society*, 36(2): 155–169.

Berten, John and Matthias Kranke (2024) 'Governing global challenges through quantified futures', *British Journal of Politics and International Relations*, 26(3): 599–621.

Bhavnani, Kum-Kum et al (eds) (2022) *Climate Futures: Reimagining Global Climate Justice*, London: Bloomsbury Publishing.

Brand, Ulrich et al (2021) 'From planetary to societal boundaries: an argument for collectively defined self-limitation', *Sustainability: Science, Practice and Policy*, 17(1): 265–292.

Buch-Hansen, Hubert and Martin B. Carstensen (2021) 'Paradigms and the political economy of ecopolitical projects: green growth and degrowth compared', *Competition & Change*, 25(3–4): 308–327.

Burke, Anthony et al (2016) 'Planet politics: a manifesto from the end of IR', *Millennium: Journal of International Studies*, 44(3): 499–523.

Buzan, Barry, Ole Wæver and Jaap de Wilde (1998) *Security: A New Framework for Analysis*, Boulder, CO: Lynne Rienner.

Cosme, Inês, Rui Santos and Daniel W. O'Neill (2017) 'Assessing the degrowth discourse: a review and analysis of academic degrowth policy proposals', *Journal of Cleaner Production*, 149: 321–334.

de Jouvenel, Bertrand (2012) *The Art of Conjecture*, Nikita Lary (trans.), Abingdon: Routledge.

'Declaration' (2008) Economic De-growth for Ecological Sustainability and Social Equity Conference, Paris, 18–19 April. Available at: https://www.degrowth.info/wp-content/uploads/2015/05/Declaration-on-Degrowth-EN.pdf

Dengler, C. and Seebacher, L.M. (2019) 'What about the Global South? Towards a feminist decolonial degrowth approach', *Ecological Economics*, 157: 246–252.

Dolez, Antoine, Céline Granjou and Séverine Louvel (2019) 'On the plurality of environmental regimes of anticipation: insights from forest science and management', *Science & Technology Studies*, 32(4): 78–96.

Dorninger, Christian et al (2021) 'Global patterns of ecologically unequal exchange: implications for sustainability in the 21st century', *Ecological Economics*, 179: 106824.

Eckersley, Robyn (2023) '(Dis)order and (in)justice in a heating world', *International Affairs*, 99(1): 101–119.

Eilstrup-Sangiovanni, Mette and Stephanie C. Hofmann (2020) 'Of the contemporary global order, crisis, and change', *Journal of European Public Policy*, 27(7): 1077–1089.

Erickson, Bruce (2020) 'Anthropocene futures: linking colonialism and environmentalism in an age of crisis', *Environment and Planning D: Society and Space*, 38(1): 111–128.

Escobar, Arturo (2015) 'Degrowth, postdevelopment, and transitions: a preliminary conversation', *Sustainability Science*, 10(3): 451–462.

Fanning, Andrew L. et al (2022) 'The social shortfall and ecological overshoot of nations', *Nature Sustainability*, 5(1): 26–36.

Fücks, Ralf (2014) 'Intelligent wachsen. Die grüne Revolution', *WSI Mitteilungen*, 67(7): 560–561.

Geels, Frank W. (2002) 'Technological transitions as evolutionary reconfiguration processes: a multi-level perspective and a case-study', *Research Policy*, 31(8–9): 1257–1274.

Geels, Frank W. (2011) 'The multi-level perspective on sustainability transitions: responses to seven criticisms', *Environmental Innovation and Societal Transitions*, 1(1): 24–40.

Gibbs, David A. and Joe Flotemersch (2019) 'How environmental futures can inform decision making: a review', *Futures*, 108: 37–52.

Granjou, Céline, Jeremy Walker and Juan F. Salazar (2017) 'The politics of anticipation: on knowing and governing environmental futures', *Futures*, 92: 5–11.

Haberl, Helmut et al (2020) 'A systematic review of the evidence on decoupling of GDP, resource use and GHG emissions, part II: synthesizing the insights', *Environmental Research Letters*, 15(6): 065003.

Hall, Peter A. (1993) 'Policy paradigms, social learning, and the state: the case of economic policymaking in Britain', *Comparative Politics*, 25(3): 275–296.

Harrington, Cameron (2016) 'The ends of the world: International Relations and the Anthropocene', *Millennium: Journal of International Studies*, 44(3): 478–498.

Hasselbalch, Jacob and Matthias Kranke (2024) 'Dealing with dangerous abundance: towards post-growth International Relations', *Review of International Studies*, 50(5): 856–865.

Hasselbalch, Jacob A., Matthias Kranke and Ekaterina Chertkovskaya (2023) 'Organizing for transformation: post-growth in International Political Economy', *Review of International Political Economy*, 30(5): 1621–1638.

Hausknost, Daniel (2020) 'The environmental state and the glass ceiling of transformation', *Environmental Politics*, 29(1): 17–37.

Hickel, Jason (2021) 'What does degrowth mean? A few points of clarification', *Globalizations*, 18(7): 1105–1111.

Hickel, Jason (2024) 'How degrowth will reshape technology', interview with Paris Marx on the podcast 'Tech Won't Save Us', Episode 226, 27 June. Available at: https://techwontsave.us/episode/226_how_degrowth_will_reshape_technology_w_jason_hickel

Hickel, Jason and Giorgos Kallis (2020) 'Is green growth possible?', *New Political Economy*, 25(4): 469–486.

Hurd, Ian (1999) 'Legitimacy and authority in international politics', *International Organization*, 53(2): 379–408.

Ikenberry, G. John (2018) 'The end of liberal international order?', *International Affairs*, 94(1): 7–23.

Lake, David A. (2013) 'Authority, coercion and power in International Relations', in Martha Finnemore and Judith Goldstein (eds) *Back to Basics: State Power in a Contemporary World*, Oxford: Oxford University Press, pp 55–77.

Lake, David A., Lisa L. Martin and Thomas Risse (2021) 'Challenges to the liberal order: reflections on *International Organization*', *International Organization*, 75(2): 225–257.

Lang, Miriam (2024) 'Degrowth, global asymmetries, and ecosocial justice: decolonial perspectives from Latin America', *Review of International Studies*, 50(5): 921–931.

Luhmann, Niklas (1976) 'The future cannot begin: temporal structures in modern society', *Social Research*, 43(1): 130–152.

'The manifesto of Degrowth Journal' (2023), *Degrowth Journal*, 1.

Mastini, Riccardo, Giorgos Kallis and Jason Hickel (2021) 'A green new deal without growth?', *Ecological Economics*, 179: 106832.

Meckling, Jonas and Bentley B. Allan (2020) 'The evolution of ideas in global climate policy', *Nature Climate Change*, 10(5): 434–438.

Newell, Peter (2024) 'Back from the dead: the ecology of IR', *International Relations*, 38(3): 331–348.

OECD (2011) *Towards Green Growth*, Paris: OECD Publishing.

Okereke, Chukwumerije (2024) 'Degrowth, green growth, and climate justice for Africa', *Review of International Studies*, 50(5): 910–920.

Pavlínek, Petr and John Pickles (2004) 'Environmental pasts/environmental futures in post-socialist Europe', *Environmental Politics*, 13(1): 237–265.

Polewsky, Max et al (2024) 'Degrowth vs. green growth: a computational review and interdisciplinary research agenda', *Ecological Economics*, 217: 108067.

Pollin, Robert (2018) 'De-growth vs a green new deal', *New Left Review*, 112: 5–25.

Purdey, Stephen J. (2010) *Economic Growth, the Environment and International Relations: The Growth Paradigm*, Abingdon: Routledge.

Raworth, Kate (2017) *Doughnut Economics: Seven Ways to Think Like a 21st-Century Economist*, London: Random House.

Schmelzer, Matthias, Andrea Vetter and Aaron Vansintjan (2022) *The Future Is Degrowth: A Guide to a World beyond Capitalism*, London: Verso.

Stoknes, Per E. and Johan Rockström (2018) 'Redefining green growth within planetary boundaries', *Energy Research & Social Science*, 44: 41–49.

Temper, Leah, Daniela del Bene and Joan Martinez-Alier (2018) 'Mapping the frontiers and front lines of global environmental justice: the EJAtlas', *Journal of Political Ecology*, 22(1): 255–278.

Tipping Point North South (2023) 'Placing the military in the degrowth narrative', *Degrowth Journal*, 1.

Vervoort, Joost and Aarti Gupta (2018) 'Anticipating climate futures in a 1.5°C era: the link between foresight and governance', *Current Opinion in Environmental Sustainability*, 31: 104–111.

Werrell, Caitlin E. and Francesco Femia (2016) 'Climate change, the erosion of state sovereignty, and world order', *Brown Journal of World Affairs*, 22(2): 221–235.

Wiedmann, Thomas et al (2020) 'Scientists' warning on affluence', *Nature Communications*, 11(1): 3107.

World Bank (2012) *Inclusive Green Growth: The Pathway to Sustainable Development*, Washington DC: World Bank.

Ziai, Aram (2015) 'Post-development concepts? Buen vivir, ubuntu and degrowth', in Boaventura de Sousa Santos and Teresa Cunha (eds) *International Colloquium Epistemologies of the South: South-South, South-North and North-South Global Learnings, Vol. 3: Other Economies*, Coimbra: Centro de Estudos Sociais, pp 143–154.

Zografos, Christos and Paul Robbins (2020) 'Green sacrifice zones, or why a green new deal cannot ignore the cost shifts of just transitions', *One Earth*, 3(5): 543–546.

8

Seeing Like a Planet: Conclusion and Pathways for International Relations Scholarship

Paul Beaumont, Lucas de Oliveira Paes and Elana Wilson Rowe

Introduction

This book's main argument has been that we must better understand how the governance of nature constrains, supports or constitutes world orders and consider with both greater scope and specificity how well our international relations (IR) concepts illuminate – or obscure – these dynamics. The attention given in the chapters to the governance of nature and how it shapes world order complements the wealth of insight that global environmental politics (GEP) scholarship has offered to better understand the politics of the environment at all scales of governance.[1] Indeed, as we argued in Chapter 1, there is not so much a 'green gap' to be filled in IR, but rather a 'green silo'. In other words, the implications of GEP scholarship have not yet been brought sufficiently to bear on IR theorizing more generally. We believe that the difficulty of influencing mainstream IR theorizing stems at least partly from the broader fragmentation of IR, which leads to scholarship that aspires to break down this silo remaining diffused across the discipline. This volume has thus sought to establish a broad tent and focal point for bringing into dialogue and amplifying the works of these scholars working at the intersection of IR and the governance of nature.

Many of the chapters take a starting point in the challenges precipitated by the interlinked crises of nature and climate that can be grouped as Anthropocene-age governance challenges. The magnitude of threats and the level of planetary change of the Anthropocene are inevitably, if belatedly, calling into question the conceptual apparatus with which IR scholars – and political practitioners – approach governance. As this volume shows, these crises undermine, threaten and, in some cases, render untenable conventional wisdom pertaining to security, modernity, peace, development and sovereignty. This is also because, as Dahlia Simangan explains in her chapter in this volume, "anthropocentric, state-centric and growth-driven characteristics of the world order are some of the underlying pathological conditions that led to this new [Anthropocene] epoch" (Chapter 2).

As several chapters in the volume document, the level of change in actual governance lags far behind what the magnitude and pace of our current crises should precipitate. Given the history of concepts traversing across the IR scholar-practitioner divide, interrogating the conceptual shortcomings and highlighting opportunities for change in terms of how IR scholars approach world ordering in the Anthropocene is both a promising scholarly direction and policy-relevant pursuit. In this concluding chapter, we draw together insights from our contributions to highlight their insights into how nature's governance shapes (or should shape) world order and the prospects for adapting – or transforming – IR concepts to better meet Anthropocene challenges.

A dialogue across diversity

Across these chapters and the different answers that they give to the questions raised in this volume, several common threads emerge that recast our understanding of the mutually constitutive relationship between the governance of nature and world order. Our contributors have diverse conceptual starting points that allow us to cover a number of core concepts in IR, such as sovereignty, security and peace, hierarchy, cooperation and modernity.

Alongside this variation in core concepts and topics covered, the chapters differ in their temporal focus, thereby ensuring that this book has reflected both historical and ongoing, as well as more emergent examples of how the governance of nature shapes world order. Yao's discussion of how the linking of nature's governance and scientific knowledge disguises power relations is a key example of the easily overlooked role that nature's governance can play in constituting the modern world order (Chapter 5). In turn, Wilson Rowe, Beaumont and Paes (Chapter 4) show that anchoring regional cooperation in border-crossing ecosystems can produce novel hierarchies that cut against dominant global power structures.

By contrast, other chapters in this book focus on how our planetary emergency could transform or is transforming such order. Bosi-Moreira and Kranke (Chapter 7) explore how the different reordering visions embedded in the projects of 'green growth' and 'degrowth' imply very different futures for multiple aspects of the world order. Meanwhile, Glaab (Chapter 6) focuses on how the growing usage of space beyond Earth challenges our understanding of both sovereignty and planetary politics.

A normative call for future change in world order commensurate to facing the challenges of the Anthropocene is called for in several of these chapters, most directly in the contributions by Simangan (Chapter 2) and McDonald (Chapter 3). These chapters foreground and problematize the human-nature distinction that underpins contemporary global governance processes. McDonald's ecological security approach would introduce nonhuman referents of security that blur that very distinction. In a similar vein, Simangan highlights the necessity of interspecies justice, addressing the vulnerability of *all* beings when crafting governance solutions for peace and security in the Anthropocene.

The methodological and empirical diversity across the chapters – as well as their varied temporal and normative frames – highlights that there are numerous scholarly and analytical pathways available for IR to meet the scale of the Anthropocene. Consequently, when we return at the end of this chapter and call for a fourth great debate, these chapters provide ballast to our claim that such a debate is not about the triumph of one theoretical school or privileging of a certain method approach over another, but rather about a mega-cognitive shift across the discipline to recognize the embeddedness of global politics in nature.

Problematizing nature's governance and world ordering

The chapters in this volume were each tasked with (re)considering key IR concepts in light of planetary governance challenges. Our starting point here is that IR's longstanding habit of bracketing nature and environment in its general theorizing is no longer tenable. Furthermore, coming to terms with and working beyond (sub)disciplinary siloes is not just an academic preoccupation. As McDonald argues in this volume, 'simply pointing to the failure of states and the broader 'international community' to address a challenge like climate change in the 40 years in which we have been aware of the problem doesn't get us to the reasons for this failure or means of addressing it. Rather, coming to terms with this failure requires examining the core assumptions and logics underpinning IR thought and global practice regarding the Anthropocene and ecological crises' (Chapter 3). Yet, as the chapters in this volume demonstrate, the conceptual transformation required is not necessarily only an agenda of innovation and new vocabulary, but also a warrant to explore understudied interconnections between processes too often studied in isolation.

Chapters 2 and 3 do indeed call for fundamental shifts in our IR conceptualizations of and in the practices of peace and security in the

Anthropocene. Both McDonald (Chapter 3) and Simangan (Chapter 2) explore in their chapters how states and global institutions have largely sought to mitigate the threats brought about by global warming, adjusting their security practices rather than rethinking security of what or for whom. Hence, McDonald argues explicitly for an 'ecological security' approach, which would encompass natural and human systems alike. Simangan illustrates the necessity of rethinking both conflict and peace as necessary for pursuing security in a world undergoing planetary-level change. Specifically, she notes the need to consider economic questions and revisit aspirations and pathways for economic growth in the pursuit of peace and security in an Anthropocene world. Likewise, in reflecting on humanity's intensifying use of outer space, Glaab invites us to rethink key binaries, such as human-nature and Earth-space. Taken together, these chapters highlight how IR's key concepts and theories are not simply external to environmental crises, but are also implicated in reproducing policy discourses and practices that they argue are conceptually incapable of addressing global environmental omni-crises. While the usual suspects – such as realism or national security ontologies – are of course cited, all three chapters identify the systematic blindspots baked into more liberal or progressive concepts too.

Other chapters illustrate how IR research can and does bring to the fore important aspects of how nature's governance shapes world order. Indeed, Wilson Rowe, Beaumont and Paes examine how transboundary ecosystem management has been entangled with the creation of novel hierarchies in the Arctic, the Amazon and the Caspian Sea regions. They demonstrate both the analytical value of hierarchy as a lens and the necessity of including hierarchies supported by 'nature' alongside the broad organizing effects of gender, civilizational and other acknowledged sources of hierarchy in world politics. Meanwhile, Yao's investigation (Chapter 5) of the role of scientific knowledge in facilitating cooperation during the International Geophysical Year highlights how the shielding of environmental cooperation from geopolitical rivalry still produced and reproduced power political effects, including the reproduction of colonial hierarchies. Similarly, approaching this question of ordering from a GEP perspective, Bosi-Moreira and Kranke (Chapter 7) note how policy practice reflects the same 'green silo' rift that this book has sought to explore and traverse. Using discursive analytical methods familiar from future studies and anticipatory governance, they document a surprising lack of conversation across the 'green growth' and degrowth/postgrowth political agendas. They highlight the divergent consequences these agendas have for future world order, as well as what it tells us about present world orders.

Taken as a whole, the chapters also reveal how the governance of nature speaks to and shapes world order by challenging, facilitating or transforming political practices relating to security, sovereignty, hierarchy, growth, cooperation and modernity. Notably, most chapters, even as they sought to interrogate one core concept, inevitably stray into one another's conceptual territory. This may be because the magnitude and interlinked nature of nature's governance – ongoing and anticipated – bring to the fore the complexity of ordering practices often disciplined along conceptual or subdisciplinary lines. The chapters in this book thus demonstrate the benefit of working across disciplinary, epistemological and methodological siloes in addressing the challenges of the governance of nature and world order.

Indeed, the chapters issue a collective warning against the longstanding tendency in IR theorizing about order to bracket nature and the environment. Several chapters demonstrate the analytical potential for IR in connecting knowledge about environmental transformations and their governance together with IR theoretical concepts on order (hierarchy, sovereignty and cooperation). Such an approach allows us to consider how there is a geopolitics of the environment in which nature and its governance are deeply embedded in and co-constituted with longstanding power structures. That geopolitics of the environment may often appear orthogonal to effective environmental governance or meeting the challenges of the Anthropocene age but are nonetheless impactful. Highlighting these structures, for example by identifying the ways in which governance of globally significant ecosystems like the Arctic and the Amazon supports and challenges regional and global hierarchies, is essential groundwork for both conceptual development and political change. It is difficult to envision how we can develop better concepts about and practices for the governance of nature and for a more just, planetary world order when we have yet to fully appreciate how implicated nature and its governance are in our current world order.

Greening international relations concepts?

This volume has also sought to encourage discussion of how IR theorizing about nature's governance and how it shapes order could develop further. As previous chapters have shown, this is not an endeavour starting from scratch; rather, it is an emerging conversation that remains disaggregated across multiple subdisciplines, including the subdiscipline of GEP that has generated much of our knowledge of how, when and why environmental cooperation emerges, succeeds and fails. This volume has thereby aimed to both complement rational-institutionalist approaches in IR and problem-solving GEP.

One way of reading these chapters then is as philosophical deconstructions of the dominant concepts within both IR (as a discipline) and 'ir' (as field of political practice). Both McDonald (Chapter 3) and Simangan (Chapter 2) show how key concepts – security and peace and conflict – are both analytically and ethically inadequate for navigating the global environmental and political crises of the Anthropocene. For McDonald, plausible alternative conceptions of security – ecological security – are in circulation, but it will require considerable 'action below and above' traditional security providers (states and defence establishments). Meanwhile, Simangan unpacks how global environmental challenges in the Anthropocene reveal the shortcomings of the anthropocentric and state-centric, foundations of the current international order, but also points out how the IR discipline has contributed to critical interrogations of this order by marshalling theoretical and methodological pluralism. As Simangan (Chapter 2) notes, the 'discipline's engagement with the Anthropocene discourse infuses a renewed sense of urgency to understand and rethink responses to the new challenges to peace and security'. Yao (Chapter 5), in her chapter exploring the dynamics of international scientific cooperation and the mastery of nature, argues convincingly that we have overlooked how the liberal international order itself is complicit in creating environmental challenges, as a driving analytical interest has been in how such an international cooperative order can solve these challenges.

Implicit in these arguments is the (reasonable and important) assumption that academic debates in IR shape and are shaped by the political practices of 'ir'. One need not search long to identify how IR has suffered 'lab leaks', of ideas such as 'clash of civilizations', that have proven unfortunately and even tragically influential (Musgrave 2021). Even more obviously, IR academics have served in influential positions in the US government (Kissinger, Nye and so on) and international organizations (for example, Ruggie). Less clearly but no less relevant, Lerner and O'Loughlin (2023) show how policy actors' own theorizing feeds back into IR's scientific ontologies (see Glaab, Chapter 6). Indeed, IR's once dominant realism paradigm can be read as a social scientific rendering of the principles of practice of realpolitik in the 19th century (Guzzini 2012), while neofunctionalist and liberal-institutionalist concepts are suffused throughout international organizations' legitimation strategies (Beaumont and Wilson Rowe 2022). In short, given that IR as a discipline is embedded and thus implicated in the 'ir' world of global politics it studies, it is not a question of whether IR influences 'ir', but of how much and through which mechanisms.

If we accept that the concepts of IR and 'ir' cross-pollinate and co-constitute one another, the stakes of this volume's collective chapters become especially apparent. The agenda to contest and transform the status quo of IR becomes part of the *political* endeavour of fostering the required transformation within world politics and the international order. Notably, in so doing, they turn IR disciplinary politics into 'actual' international politics. In particular, these chapters contest the conventional wisdom that if IR's green gap can be closed by growing GEP, then the rest of IR can carry on as before. Indeed, in different ways, McDonald and Simangan illustrate how

IR's conceptual apparatus is co-constitutive and partly therefore implicated in enabling the omni-crises of the Anthropocene. Zooming out even further, Glaab's chapter shows that contemporary space activities (discourses and practices) are co-constituted by existing understandings of sovereignty. She argues that this understanding of sovereignty – and associated practices of governing from this understanding – means that we risk repeating in space the colonial and extractivist pathologies of terrestrial human history. These chapters suggest and argue for a thoroughgoing conceptual transformation of IR far beyond the GEP subfield. Hence, these works offer an extended explication of Corry and Stevenson's warning against simply adding 'environment to existing theoretical frameworks for understanding global politics – and conversely adding "governance" to natural science analysis of Earth Systems' (Corry and Stevenson 2018: 194).

As well as problematizing IR's key concepts, the chapters in this volume also highlight the necessity for IR to self-consciously foster theoretical and methodological pluralism in order to reckon with the challenges ushered in by the Anthropocene (see Beaumont and Coning 2022). Indeed, several of the contributors highlight how the epochal scale and complexity of the transformations and crises ahead will require systematic means of thinking about the future. In the spirit of innovating in the face of urgency and taking a broad understanding of science as the 'systematic production of knowledge' (Jackson 2010), our volume's contributors showcase a number of approaches that traverse sustained philosophical reflection, discourse analysis and macro-history. Bosi-Moreira and Kranke in Chapter 7 go furthest in explicitly theorizing how we can empirically study *futures* through the concept of *imaginaries*. As the authors explain: 'Imaginaries of certain futures, rather than others, affect the contemporary realm of possibility, as neatly captured in the term "present future"'. They go on to explore how global sustainability governance constructs environmental issues (or nature) 'with reference to not only their present constitution but also their anticipated future constitution'(Bosi-Moreira and Kranke, Chapter 7). Unpacking the different futures envisioned through the degrowth and postgrowth policy agendas, they show how despite sharing a belief in the need for rapid transformation, each holds radically different implications for the extent to which the international order – the status quo – requires overhauling. While McDonald, Simangan and Glaab also showcase how conceptual analysis of *present* discourses can shed light on the potentialities of the present and the risks of the future, the contributions by Yao and Wilson, Beaumont and Paes develop historical/processual analyses that illuminate how efforts to govern nature recursively shape and even constitute the actors doing the governing in the first place.[2]

Ultimately, we suggest that (1) normative-cum-empirical critiques and (2) longitudinal/historical analyses provide promising analytical pathways for how to go about reversing the siloing of nature's governance. While relying on quite different meta-theoretical foundations, each engages in a sustained attempt to encourage researchers, policy makers and citizens to break with anthropocentric habits either via critical reflection on the concepts scholars and policy makers think with and thus act through, and/or via careful, theoretically informed empirical exploration into how nature's governance has shaped and is shaping world order in an ongoing fashion. While this volume has illustrated a number of plausible means of conducting such analysis, we would encourage further sustained attention to the question of how to systematically study nature-governance co-constitution.

Towards a fourth debate: reckoning with the 'environment' of global politics

It is often said that climate change (and we would add other interlinked ecological crises) is no longer a scientific problem, but a political one. Many of the chapters in this volume are premised on how this political challenge is not limited to the policy world, but must engage the status quo within academia, and specifically the discipline of IR too. Indeed, the primary motivating force behind this volume was the sense that, despite both the growth of the GEP subdiscipline and the rise of calls for transforming IR in light of the Anthropocene, the potential impact of this scholarship has been dampened by its diffusion across IR's subfields and campfires, and amid its relentless theoretical turn taking (Sylvester 2013; Heiskanen and Beaumont 2024). The fragmentation of IR has thus been both a blessing and a curse in terms of putting environmental matters on the agenda in IR. The blessing has been that GEP and critical work on the environment have been able to develop, thrive and escape the gatekeeping practices of major journals that might have inhibited both the amount and substance of work. The curse is that the opportunity to flourish outside the mainstream reduces the incentives to engage with the mainstream and even with fellow travellers. In bringing together a selection of this work under one roof, we hope to contribute towards remedying this issue by providing a focal point for dialogue, cross-fertilization and inspiration, and use the collective enterprise to amplify its individual contributions.

Reflecting upon the chapters and their points of dialogue, we hope that this volume may contribute to the discipline beyond expanding the IR 'green silo'. Across these quite different approaches in each chapter, there is a common commitment that could serve as an anchor for future research and perhaps even the makings of an overdue 'great debate' concerning planetary change and IR. Taking stock of IR's so-called 'great debates' over time, it is possible to discern a steady movement away from first-order ('real world') problems.[3] Indeed, in IR's last such debate, the positivist mainstream of IR underwent sustained epistemological and methodological critique from feminists, poststructuralists and postcolonial scholarship. While transformative for IR, this third great debate pertained almost exclusively to meta-theory and what constituted the appropriate conduct of inquiry for IR. The downstream consequence was that 'critical IR' developed as a subdiscipline and enabled a pragmatic but unsatisfying 'theoretical peace' to emerge that was based on systematic disengagement with one another (Dunne, Hansen and Wight 2013; Jahn 2021; Heiskanen and Beaumont 2024).

We believe that this volume shows that this theoretical peace requires breaching: the global environmental omni-crises and the required revolution in governance provide an imperative for a 'fourth debate' that returns IR to first-order questions and requires systematic re-engagement across 'mainstream IR' and its subdisciplines. We hope that this volume will contribute to identifying the contours of what such a debate may involve: the systematic re-appraisal of core IR concepts' analytical and ethical adequacy in an epoch when the mutual constitution of global politics and its 'environment' can no longer be ignored. But calling for such a debate is not enough; it will require a sustained and concerted effort on behalf of IR scholars and IR's institutions to enable such a conversation. We are optimistic that IR has the intellectual resources to meet the challenge. Still, it remains an open question whether the structural fragmentation of the field and the status quo-reproducing forces will prove too strong.

Notes

[1] See Liftin 1998, 2000; Adger et al 2001; Corry 2013; Green 2013; McDonald 2013, 2021; Dalby 2014, 2020; Beck, Esguerra and Goerg 2017; Corry and Stevenson 2017; Green and Hale 2017; Aykut, Foyer and Morena 2018; Glaab and Fuchs 2018; Yao 2019, 2021, 2022; Colgan, Hale, and Green 2021; Wilson Rowe 2021; Aykut and Maertens 2022; Paes 2022, 2023; Beaumont and Wilson Rowe 2022; Maglia and Wilson Rowe 2023, 2024.

[2] In a sense, these chapters complement the research agenda exploring the 'globalisation of climate, and the climatization of global issues' (Aykut, Foyer and Morena 2018, 4). *Climatization* illuminates how climate change discourse, and its associated logics is gradually being incorporated and transforming policy fields 'formerly unrelated to the climate problem'.

[3] The usual narrative of IR runs that realists and liberals debated the prospects of peace during the interwar period, which was settled in the former's favour with the outbreak of the Second World War. The second great debate concerned the optimum methods of IR and pitted qualitative scholars, led by Hedley Bull, against the rising behaviouralist movement, led by Morton Kaplan. See Wæver (1998) for a famous account of IR's great debates.

References

Adger, Neil, Tor Arve Benjaminsen, Katrina Brown and Hanne Svarstad (2001) 'Advancing a political ecology of global environmental discourses'. *Development and Change*, 32(4), 681–715.

Aykut, Stefan C. and Maertens, Lisa (2022) 'The climatization of global politics: introduction to the special issue', in *The Climatization of Global Politics*. Cham: Springer International Publishing, pp 1–18.

Aykut, Stefan C., Jean Foyer and Édouard Morena, eds (2018) *Globalising the Climate: COP21 and the Climatisation of Global Debates*. New York: Routledge.

Beck, Silke, Alejandro Esguerra and Christoph Goerg (2017) 'The co-production of scale and power: the case of the millennium ecosystem assessment and the intergovernmental platform on biodiversity and ecosystem services'. *Journal of Environmental Policy & Planning*, 19(5), 534–549.

Beaumont, Paul and Cedric de Coning (2022) 'Coping with complexity: toward epistemological pluralism in climate–conflict scholarship'. *International Studies Review* 24(4), viac055.

Beaumont, Paul and Elana Wilson Rowe (2022) 'Space, nature and hierarchy: the ecosystemic politics of the Caspian Sea'. *European Journal of International Relations*, December. https://doi.org/10.1177/13540661221142179

Colgan, Jeff D., Jessica F. Green and Thomas N. Hale (2021) 'Asset revaluation and the existential politics of climate change'. *International Organization*, 75(2), 586–610.

Corry, Olaf (2013) *Constructing a Global Polity*. London: Palgrave Macmillan.

Corry, Olaf and Hayley Stevenson, eds (2018) *Traditions and Trends in Global Environmental Politics: International Relations and the Earth*. New York: Routledge.

Dalby, Simon (2014) 'Environmental geopolitics in the twenty-first century'. *Alternatives*, 39(1), 3–16.

Dalby, Simon (2020) *Anthropocene Geopolitics: Globalization, Security, Sustainability*. Ottawa: University of Ottawa Press.

Dunne, Tim, Lene Hansen and Colin Wight (2013) 'The end of international relations theory?'. *European Journal of International Relations*, 19(3), 405–425.

Glaab, Katharina and Doris Fuchs (2018) 'Green faith? The role of faith-based actors in global sustainable development discourse'. *Environmental Values*, 27(3), 289–312.

Green, Jessica F. (2013) *Rethinking Private Authority: Agents and entrepreneurs in global environmental governance*. Princeton: Princeton University Press.

Green, Jessica F. and Thomas N. Hale (2017) 'Reversing the marginalization of global environmental politics in international relations: an opportunity for the discipline'. *PS: Political Science & Politics*, 50(2), 473–479.

Guzzini, Stefano, ed. (2012) *The Return of Geopolitics in Europe? Social Mechanisms and Foreign Policy Identity Crises*. Cambridge: Cambridge University Press.

Heiskanen, Jaakko and Paul Beaumont (2024) 'Reflex to turn: the rise of turn-talk in international relations'. *European Journal of International Relations*, 30(1), 3–26.

Jackson, Patrick Thaddeus (2010) *The Conduct of Inquiry in International Relations*. Abingdon: Routledge.

Jahn, Beate (2021) 'Critical theory in crisis? a reconsideration'. *European Journal of International Relations*, 27(4), 1274–1299.

Lerner, Adam B. and Ben O'Loughlin (2023) 'Strategic ontologies: narrative and meso-level theorizing in international politics'. *International Studies Quarterly*, 67(3), sqad058.

Litfin, Karen, ed. (1998) *The Greening of Sovereignty in World Politics*. Cambridge, MA: MIT Press.

Litfin, Karen (2000) 'Environment, wealth, and authority: global climate change and emerging modes of legitimation'. *International Studies Review*, 2(2), 119–148.

Maglia, Cristiana and Elana Wilson Rowe (2023) 'Ecosystems and ordering: exploring the extent and diversity of ecosystem governance'. *Global Studies Quarterly*, 3(2), ksad028.

Maglia, Cristiana and Elana Wilson Rowe (2024) 'Ecosystems and ordering: a dataset'. *Data in Brief*: 111085.

McDonald, Matt (2013) 'Discourses of climate security'. *Political Geography*, 33: 42–51.

McDonald, Matt (2021) *Ecological Security*. New York: Cambridge University Press.

Musgrave, Paul (2021) 'Political science has its own lab leaks'. *Foreign Affairs*, 3. Available at: https://foreignpolicy.com/2021/07/03/political-science-dangerous-lab-leaks/

Paes, Lucas de Oliveira (2022) 'The Amazon rainforest and the global–regional politics of ecosystem governance'. *International Affairs*, 98(6), 2077–2097.

Paes, Lucas De Oliveira (2023) 'Networked territoriality: a processual–relational view on the making (and makings) of regions in world politics'. *Review of International Studies*, 49(1), 53–82.

Sylvester, Christine (2013) 'Experiencing the end and afterlives of international relations/theory'. *European Journal of International Relations*, 19(3), 609–626.

Wæver, Ole (1998) 'The sociology of a not so international discipline: American and European developments in international relations'. *International Organization*, 52(4), 687–727.

Wilson Rowe, Elana (2021) 'Ecosystemic politics: analyzing the consequences of speaking for adjacent nature on the global stage'. *Political Geography*, 91: 102497.

Yao, Joanne (2019) '"Conquest from barbarism": the Danube Commission, international order and the control of nature as a standard of civilization'. *European Journal of International Relations*, 25(2), 335–359.

Yao, Joanne (2021) 'An international hierarchy of science: conquest, cooperation, and the 1959 Antarctic Treaty System'. *European Journal of International Relations*, 27(4), 995–1019.

Yao, Joanne (2022) *The Ideal River: How Control of Nature Shaped the International Order.* Manchester: Manchester University Press.

Index

References to figures appear in *italic type*; those in **bold type** refer to tables.
References to endnotes show both the page number and the note number (57n5).

A

absolute gains 91
Amazon 69, 76–81
Amazon Cooperation Treaty (ACT) 69–70, 75, 76, 78–79, 79–80
Amazon Cooperation Treaty Organization (ACTO) 79
animal ethics 29
Antarctic Treaty System (ATS) 101, 116
Anthropocene 6, 7–8, 20–21, 22, 48, 49, 148
  as paradigm shift in international relations 26–27
  peace and security in **22**, 23, 27–32
Anthropocene security 53
anthropocentrism **22**, 23, 25
Arctic 65, 70–76, 80–81
Arctic Council 70–71, 73–75, 81
Arctic Ocean 70, 72
Artemis Accords 118
Assault on the Unknown (Sullivan) 99–100, 105
Astronomical Unit 93–94
astropolitics 114
Australia 51, 94, 97, 103, 104–105

B

Banks, J. 104
border-crossing ecosystems 62–63
  governance 64–65, 81–83
    Amazon 69, 76–81
    Arctic 65, 70–76, 80–81
    Caspian Sea 65, 67–70, 71, 75, 80
  and hierarchies 65–67
bounded territoriality 26
Brazil 69–70, 78

C

Canada 1, 97
capitalocene 49
Caspian Sea 65, 67–70, 71, 75, 80

Chapman, S. 99
China 50, 75, 117
civilizational progress 104–106
climate change 2, 25, 91
  definitions 57n5
  and security 27–28, 42–44, 45, 46–47, 48, 49–52
  ecological security 53–55, 56
  and territoriality 27
  UN New Agenda for Peace 32–33
climate financing 24–25
climate justice 136
Clinton, H. 101
Cold War 12, 30, 95, 97, 114, 117
colonialism 90, 102–104, 115
complexity theory 27
Conference of Parties (COP) 2023 33
Cook, J. 94, 104
critical environmental scholarship 6–8

D

data gathering 92–93, 94, 95, 102–104
decoupling 131
deep hierarchies 66
degrowth 127–128, 129–130, 130–135
  implications for international relations (IR) 135–137
Degrowth Conference 133
Degrowth Journal 133–134
Delisle, J.-N. 93
Dolman, E. 114, 116
Doomsday Clock 33
doughnut economies 134
Dunbar, J. 90

E

Earth orbit 120–121
Earth-space sustainability 121
Earth System Governance project 4–5
Earthrise era 122

ecological challenges *see* environmental challenges
ecological security 53–55, 56
economic growth 31–32
economic power 135–136
ecosystem interdependence 69
ecosystems 62–63
  and ordering 63–64
  transboundary governance 64–65, 81–83
    Amazon 69, 76–81
    Arctic 65, 70–76, 80–81
    Caspian Sea 65, 67–70, 71, 75, 80
    and hierarchies 65–67
Enlightenment 63–64
environmental agreements 64
environmental challenges 2
  and security 27–29, 47–48
  see also climate change
environmental futures 127–128
  green growth and degrowth 130–135
    implications for international relations (IR) 135–137
  and world (re)ordering 128–130
environmental justice 136
environmental politics *see* global environmental politics (GEP)
epistemic completion 88–89, 92–95, 101–102
  International Geophysical Year (IGY) 95–101
Essays on the History of Mankind (Dunbar) 90
ethics of war 29
Eurocene 48–49

F

forest carbon density, Amazon 77
future *see* environmental futures

G

geopolitics, outer space as 113, 114–116
global commons, outer space as 116–120
global environmental politics (GEP) 3–6, 8, 89, 91, 147, 150
  see also international cooperation; international relations (IR)
global knowledge *see* epistemic completion
governance of nature 2, 21, 90–92, 145–147
  transboundary ecosystems 64–65, 81–83
    Amazon 69, 76–81
    Arctic 65, 70–76, 80–81
    Caspian Sea 65, 67–70, 71, 75, 80
    hierarchies / hierarchical relationships 65–67
  see also global environmental politics (GEP)
green growth 127–128, 129–130, 130–135
  implications for international relations (IR) 135–137
Green New Deals 135

growth *see* economic growth; green growth
Guterres, A. 30, 32

H

Halley, E. 93
hierarchies / hierarchical relationships 65–67
  Amazon 69, 76–81
  Arctic 65, 70–76, 80–81
  Caspian Sea 67–70, 71, 75, 80
  International Geophysical Year (IGY) 101–106
Holocene conditions 6
human-nature entanglement 29
human security 23

I

Inclusive Green Growth (World Bank) 133
India 50, 97
Indigenous Peoples 121
Indigenous thinking 27
Intergovernmental Panel on Climate Change (IPCC) 33, 48
international community 45–46
international cooperation 88, 89–92
  epistemic completion 92–95
  see also global environmental politics (GEP); International Geophysical Year (IGY)
international environmental agreements 64
International Geophysical Year (IGY) 88, 94, 95–101
  and global hierarchies 101–106
international relations (IR) 2–3, 4, 5, 7, 8, 150–151
  academic debates in 148–149
  and Anthropocene discourse 21, 26–27
  implications of green growth and degrowth 135–137
  and linearity 25
  outer space politics 113
  posthumanist turn 23
  security in 44–47
  see also global environmental politics (GEP)
international security framework 45–46, 47
international society 46
interspecies togetherness 29

J

justice 136

K

Kant, I. 89–90
Kármán line 119
Keohane, R. 3
knowledge, global *see* epistemic completion

L

linearity **22**, 25
Litfin, K.T. 121
Locke, J. 90

INDEX

M

Mann, M. 29
Marlowe, C. 98–99
Mearsheimer, J. 91
meteorology 92–93, 100
militarism 29–30
Moon Agreement 117
Morgenthau, H. 89
multilevel perspective (MLP) 134

N

national security framework 45, 46, 54
nature
 control of 62, 89–92
 governance of 2, 21, 64–65, 81–83, 145–147 (*see also* global environmental politics (GEP))
  Amazon 69, 76–81
  Arctic 65, 70–76, 80–81
  Caspian Sea 65, 67–70, 71, 75, 80
  hierarchies / hierarchical relationships 65–67
 human-nature entanglement 29
 ordering of 63–64
Navajo Nation 121
North Water Polynya 1
nuclear weapons 30, 33

O

Odishaw, H. 96, 100, 105
orbital space 120–121
order 129–130
ordering of nature 63–64
Organisation for Economic Cooperation and Development (OECD) 132–133
outer space 112–113
 as geopolitics 114–116
 as global commons 116–120
outer space governance 119, 120–123
Outer Space Treaty (OST) 101, 117, 118, 120, 123

P

peace and security
 in the Anthropocene **22**, 23, 27–32
 UN New Agenda for Peace 32–34
 see also security
Philip, Duke of Edinburgh 96
Pikialasorsuaq 1
Pikialasorsuaq Commission 1
Pingré, A. 94
Pituffik Space Base 1
planetary boundaries 6, 47–48
planetary politics 121–122
 see also post-planetary politics
plastic input, Arctic Ocean 72
Politics among Nations (Morgenthau) 89
post-planetary politics 122–123

postage stamps 99, *100*
postgrowth *see* degrowth
power 135–136
present future 127

R

reconciliation mechanisms 29
regionalism 78, 79
relationality 27, 30–31
relative gains 91
Remarks on the Present State of Meteorological Science (Ruskin) 92–93
Restless Sphere, The (BBC) 96, 98
right-sizing 134
Ruggie, J.G. 26
Ruskin, J. 92–93, 94
Russia 24, 50, 67, 70, 75, 117
 see also Soviet Union

S

security 23, 42, 68, 136, 147–148
 in the Anthropocene **22**, 23, 27–32
 and climate change 42–44, 45, 46–47, 48, 49–52
 ecological security 53–55, 56
 and environmental challenges 47–48
 in international relations (IR) 44–47
 traditional accounts of 52, 55–56
 see also peace and security
Shelton, J.P. 97
social contract 44
solar radiation management (SRM) 23
sovereignty 114–115, 118–119, 135
Soviet Union 67, 97, 103, 114
 see also Russia
space *see* orbital space; outer space
space exploration 115–116
Space Liability Convention 120
space race 112, 114
state-centrism **22**, 24–25
states
 and security 44–45, 46, 49–50, 50–51
 ecological security 54
Sullivan, W. 96–97, 99–100, 105
sustainability 121

T

Tamburlaine the Great (Marlowe) 98–99
technological innovation 131
Tehran Convention 68, 69, 75
territoriality 25, 26–27
Third Nuclear Age 30
Thule Air Base 1
Towards Green Growth (OECD) 132–133
transboundary ecosystems 62–63
 governance 64–65, 81–83
  Amazon 69, 76–81
  Arctic 65, 67–70, 80–81

Caspian Sea 65, 67–70, 71, 75, 80
and hierarchies 65–67

U

UN Charter 42, 57n4
UN Committee on the Peaceful Uses of Outer Space (COPUOS) 117
UN Development Programme (UNDP) 20
UN Educational, Scientific and Cultural Organization (UNESCO) 78, 96, 100
UN Framework Convention on Climate Change (UNFCC) 12, 33, 49–50
UN New Agenda for Peace 32–34
UN Security Council (UNSC) 43, 47, 50, 56
unbundled territoriality 26
UNESCO Courier 94, 99
United Nations (UN) 24, 28, 32
United States (US) 97, 114, 115
postage stamps 99, *100*

V

Venus 93–94, 104
vulnerability 28–29

W

war 30
see also Cold War
war ethics 29
waste trade 31
World Bank 133
world (re)ordering 128–130
green growth and degrowth 130–135
worldviews / world order 22–26, **22**

www.ingramcontent.com/pod-product-compliance
Lightning Source LLC
Chambersburg PA
CBHW071711020426
42333CB00017B/2215